Get Started
Astronomy

Get Started

Astronomy

DK

LONDON, NEW YORK, MUNICH, MELBOURNE, DELHI

Senior Editor Helen Fewster
Senior Art Editor Alison Shackleton
Managing Editor Penny Warren
Picture Researcher Sarah Hopper
Senior Jacket Creative Nicola Powling
Jacket Designer Rosie Levine
Senior Pre-production Producer Tony Phipps
Senior Producer Alex Bell
Art Director Jane Bull
Publisher Mary Ling

DK Publishing
North American Consultant Christina Pease
Editors Jill Hamilton, Allison Singer

DK India
Editor Ligi John
Assistant Editor Neha Ruth Samuel
Senior Art Editor Ira Sharma
Art Editor Anjan Dey
Assistant Art Editor Pallavi Kapur
Managing Editor Alicia Ingty
Managing Art Editor Navidita Thapa
Pre-production Manager Sunil Sharma
DTP Designers Rajdeep Singh, Neeraj Bhatia

Written by Robert Dinwiddie

This edition first published in 2013 by DK Publishing,
4th floor, 345 Hudson Street, New York, New York 10014

15 16 17 18 10 9 8 7 6 5 4
009-196196-Dec/2013

Published in Great Britain by Dorling Kindersley Limited.

A catalog record for this book is available from the Library of Congress.

ISBN 978-1-4654-1584-4

DK books are available at special discounts when purchased in bulk for
sales promotions, premiums, fund-raising, or educational use. For details,
contact: DK Publishing Special Markets, 345 Hudson Street, New York,
New York 10014 or SpecialSales@dk.com.

Printed and bound in China

Discover more at
www.dk.com

Contents

1
Start Simple

2
Build On It

3
Take It Further

PUBLISHER'S NOTE
Looking at the Sun with the naked eye,
binoculars, or a telescope can cause eye
damage. Advice on safe viewing of the Sun
is included on pages 160 and 161 of this book.
The author and the publishers cannot accept
any liability to readers failing to follow this
advice. Modifying cameras or other equipment
may invalidate the manufacturer's warranty
and readers do so at their own risk.

Build Your Course

This course is divided into three parts. Start Simple introduces the night sky, focusing on objects you can see with the naked eye. Build On It explains what you can see with binoculars, while in Take It Further you'll learn about objects best viewed through a telescope.

Observation techniques and activities

Within each chapter you'll find information on particular astronomical objects or phenomena to look for in the night sky—such as the Moon, a specific planet, a bright comet, an eclipse, or a particular deep sky object. You'll also find advice on when and where you will be able to see these specific objects or events, as well as what to look out for when viewing them, and tips on observation technique and how to navigate your way across the sky at night.

Key constellations

In each chapter, you'll also find many double-page features that focus on key constellations. Each includes a figure and star map for the constellation, with labels indicating the more interesting objects to view in it, and "locator" diagrams to help you work out when and where to observe the constellation in the night sky. Some of the constellations covered are visible to viewers in one hemisphere only (Northern or Southern), but many can be seen by viewers in either hemisphere.

Taurus constellation figure

Taurus locator diagram

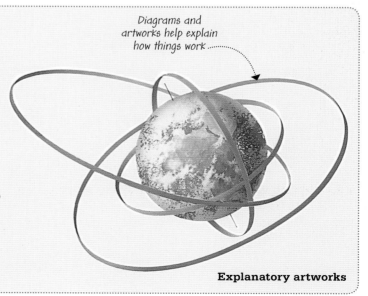

Crab Nebula in Taurus

How does it work?

In addition to the various observation exercises and constellation-based features, look out for background information —as well as concise explanations on how various astronomical phenomena work—in the "Quick Guides" scattered throughout the book. For example, there are quick guides to all the bright planets, the Sun and Moon, and to stars, comets, and the Milky Way Galaxy. You'll also find plenty of explanatory diagrams, covering everything from satellite orbits to the different types of solar eclipse.

Diagrams and artworks help explain how things work

Explanatory artworks

Our Place in Space

Our planet is traveling through a boundless, three-dimensional entity called "space." This is almost empty, but at varying distances from Earth vast numbers of other concentrations of matter—planets, stars, and galaxies—also exist. These objects, and the emptiness separating them, are known as "the Universe." Astronomy is the study of these objects.

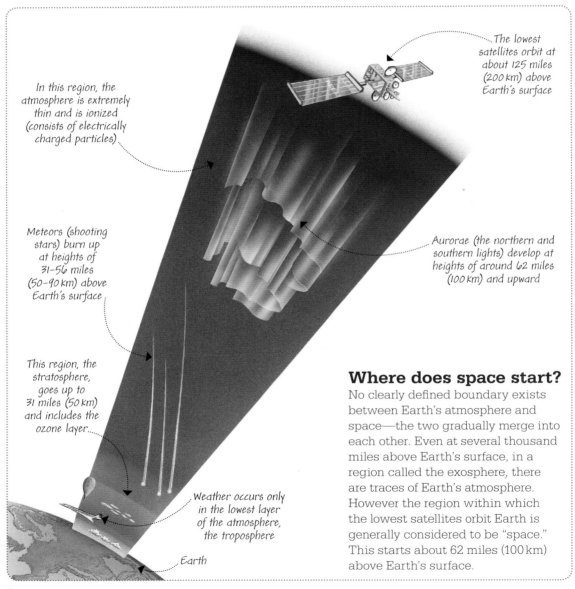

...The lowest satellites orbit at about 125 miles (200 km) above Earth's surface

In this region, the atmosphere is extremely thin and is ionized (consists of electrically charged particles)...

Aurorae (the northern and southern lights) develop at heights of around 62 miles (100 km) and upward

Meteors (shooting stars) burn up at heights of 31–56 miles (50–90 km) above Earth's surface

This region, the stratosphere, goes up to 31 miles (50 km) and includes the ozone layer...

Weather occurs only in the lowest layer of the atmosphere, the troposphere

Earth

Where does space start?

No clearly defined boundary exists between Earth's atmosphere and space—the two gradually merge into each other. Even at several thousand miles above Earth's surface, in a region called the exosphere, there are traces of Earth's atmosphere. However the region within which the lowest satellites orbit Earth is generally considered to be "space." This starts about 62 miles (100 km) above Earth's surface.

The big picture

The Universe is vast—in fact it is possibly infinite in size—and everything in it is part of something bigger. One way to get a feel for the size of the Universe is to use a series of astronomical stepping stones, as shown here—from our planet, to the Solar System that contains our planet, to the Milky Way Galaxy, and so on. Some of the largest known structures in the Universe are strings of galaxy superclusters, which extend for billions of light-years through space.

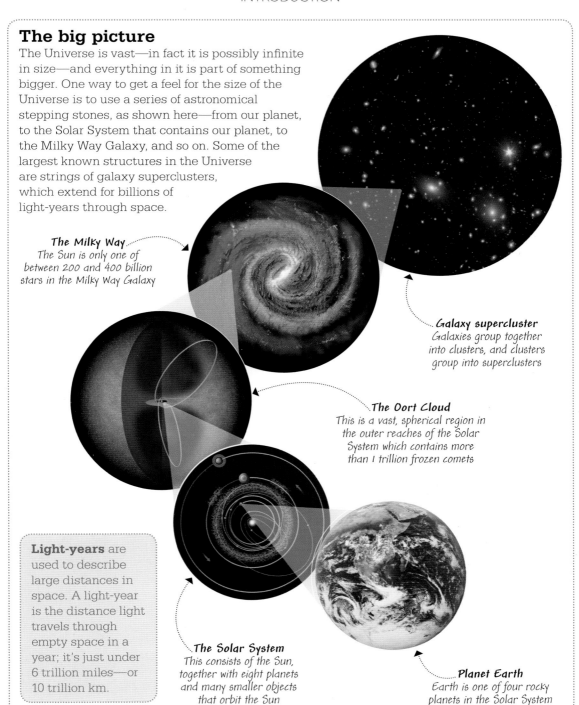

The Milky Way
The Sun is only one of between 200 and 400 billion stars in the Milky Way Galaxy

Galaxy supercluster
Galaxies group together into clusters, and clusters group into superclusters

The Oort Cloud
This is a vast, spherical region in the outer reaches of the Solar System which contains more than 1 trillion frozen comets

Light-years are used to describe large distances in space. A light-year is the distance light travels through empty space in a year; it's just under 6 trillion miles—or 10 trillion km.

The Solar System
This consists of the Sun, together with eight planets and many smaller objects that orbit the Sun

Planet Earth
Earth is one of four rocky planets in the Solar System

Our Place in the Universe

Our neighborhood in space is the Solar System. This consists of the Sun, together with eight planets and many smaller objects—such as asteroids, comets, and planetary moons—that orbit the Sun.

The Solar System

Objects in the Solar System fall into different groups, which generally occupy different regions. Four rocky planets (Mercury, Venus, Earth, and Mars) are closest to the Sun. Beyond Mars is the asteroid belt, where small rocky asteroids orbit. Next is a region in which four much larger planets—"gas giants" Jupiter, Saturn, Uranus, and Neptune—move around. Beyond Neptune lie two vast areas occupied by comets and other small objects made of rock and ice: the Kuiper belt, and farther out, the Oort cloud.

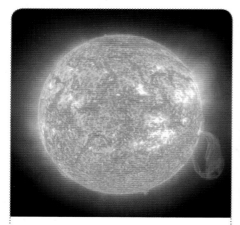

The Sun our closest star

The Sun is a medium-sized star—a huge ball of hot, luminous gas, which produces energy from nuclear fusion reactions in its core (central region). It has existed in this form for some 4.6 billion years and should stay the same for another 5 billion years or so.

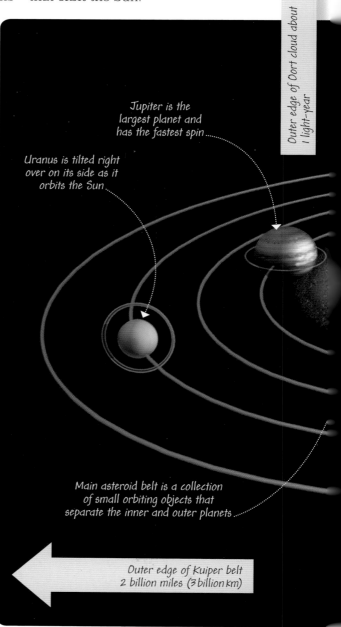

Jupiter is the largest planet and has the fastest spin

Uranus is tilted right over on its side as it orbits the Sun

Outer edge of Oort cloud about 1 light-year

Main asteroid belt is a collection of small orbiting objects that separate the inner and outer planets

Outer edge of Kuiper belt 2 billion miles (3 billion km)

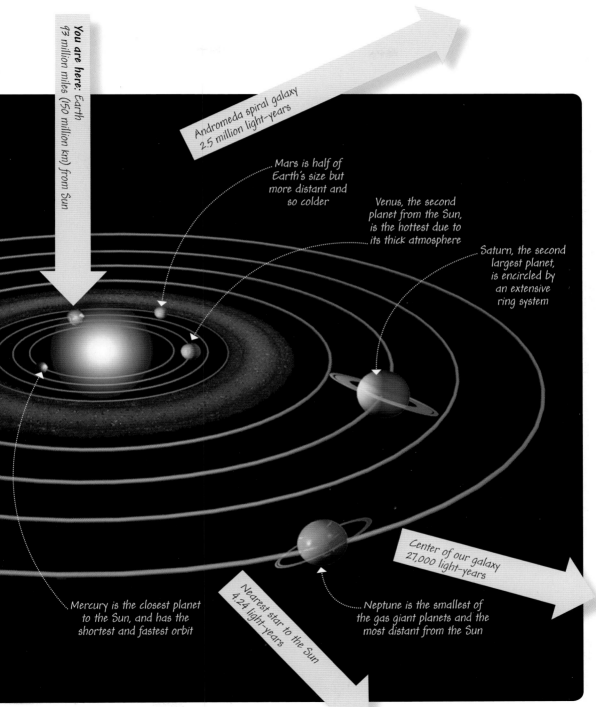

You are here: Earth 93 million miles (150 million km) from Sun

Andromeda spiral galaxy 2.5 million light-years

Mars is half of Earth's size but more distant and so colder

Venus, the second planet from the Sun, is the hottest due to its thick atmosphere

Saturn, the second largest planet, is encircled by an extensive ring system

Center of our galaxy 27,000 light-years

Mercury is the closest planet to the Sun, and has the shortest and fastest orbit

Nearest star to the Sun 4.24 light-years

Neptune is the smallest of the gas giant planets and the most distant from the Sun

Earth and its Closest Neighbors

The Solar System's four inner planets—Earth, Mars, Venus, and Mercury—look very different on the surface, but they were formed from the same materials and share many features in common. All four have a central iron core surrounded by a mantle and crust of silicon-rich rock.

The rocky planets

The inner planets are often described as "rocky" or "terrestrial" because they formed from rock and metal. As they developed, heavier materials like metal sank to the core, and lightweight rock rose to the surface. Sun-scorched Mercury has almost no atmosphere, but Venus and Mars are covered by a blanket of carbon dioxide. Unlike the others, Earth is the only planet with oxygen-rich air and liquid water on the surface.

Size matters Over time planets and objects roving across the sky appear to change in size. This apparent size, or angular diameter, is measured in degrees, arcminutes, or arcseconds: 1° is 60 arcminutes; 1 arcminute is 60 arcseconds. A closed fist held at arm's length is approximately 10°.

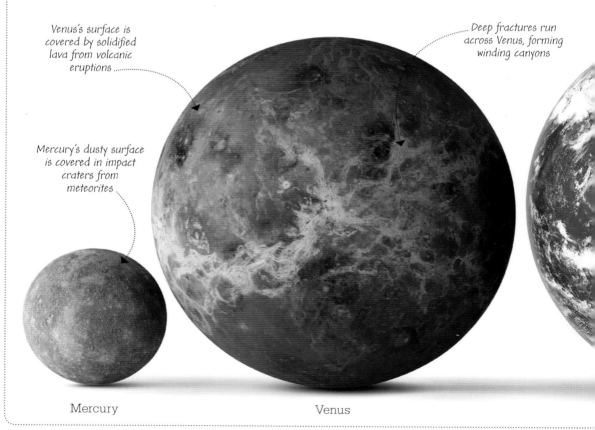

Venus's surface is covered by solidified lava from volcanic eruptions ......

Deep fractures run across Venus, forming winding canyons

Mercury's dusty surface is covered in impact craters from meteorites ......

Mercury

Venus

Our natural satellite

Earth is the only rocky planet with a large natural satellite. The Moon formed around 4.5 billion years ago, coalescing from a cloud of debris after Earth was struck by a minor planet. The dark "seas," or maria, on the Moon's surface are plains of volcanic rock that formed when large impacts fractured the Moon's crust, releasing vast floods of lava onto the surface. Surrounding the sea are paler highlands covered in ancient craters. Like the rocky planets, the Moon is thought to have a partially molten core of iron. It is the brightest object in the sky after the Sun.

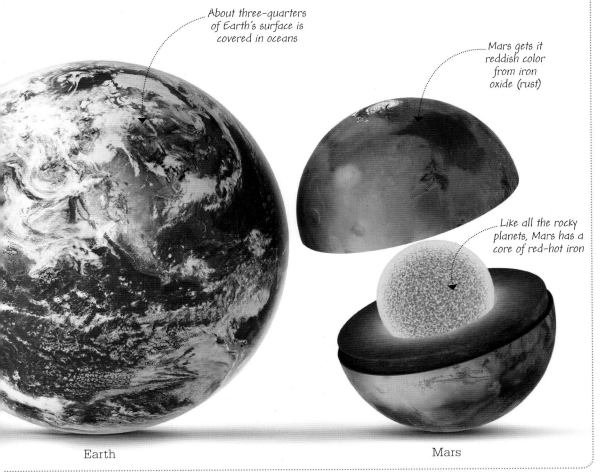

About three-quarters of Earth's surface is covered in oceans

Mars gets it reddish color from iron oxide (rust)

Like all the rocky planets, Mars has a core of red-hot iron

Earth

Mars

Gas Giants and Asteroids

Beyond the main asteroid belt lie four gigantic planets: gas giants Jupiter, Neptune, Saturn, and Uranus. Very different from the rocky inner planets, these strange worlds are vast gas and liquid globes composed of substances such as hydrogen, helium, and methane.

The gas giant planets

Despite the name, the gas giants are mostly liquid and have small solid cores. All have a deep, stormy atmosphere and rings made of rock or ice fragments. Mighty Jupiter is the largest planet by far, its mass more than twice that of all other planets combined. Like Saturn, it spins so fast that its surface clouds have stretched out into bands.

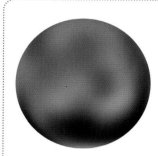

Remember Pluto?

When Pluto was discovered in 1930, it was considered to be the ninth planet. In recent years astronomers have found numerous Pluto-like objects in the outer Solar System— some larger than Pluto. As a result, Pluto was reclassified in 2006 as a dwarf planet.

Jupiter's cloudy outer atmosphere is 620 miles (1,000 km) thick

Deep inside Jupiter, intense pressure turns hydrogen into a liquid metal

A layer of liquid hydrogen and helium lies under the outer atmosphere

The Great Red Spot is a storm that has been raging for more than 300 years

Jupiter

Neptune

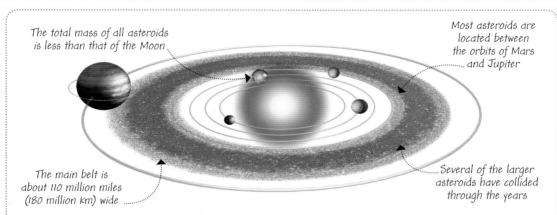

The total mass of all asteroids is less than that of the Moon

Most asteroids are located between the orbits of Mars and Jupiter

The main belt is about 110 million miles (180 million km) wide

Several of the larger asteroids have collided through the years

The asteroid belt

Asteroids are giant rocks that drift around the inner Solar System. Most lie in a belt between Mars and Jupiter, but some occasionally come dangerously close to Earth. The smallest are the size of houses, while the very largest is also classified as a dwarf planet.

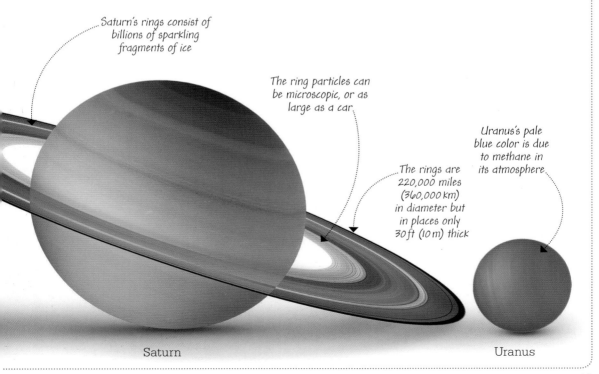

Saturn's rings consist of billions of sparkling fragments of ice

The ring particles can be microscopic, or as large as a car

The rings are 220,000 miles (360,000 km) in diameter but in places only 30 ft (10 m) thick

Uranus's pale blue color is due to methane in its atmosphere

Saturn

Uranus

What Can You See?

With or without binoculars, there's plenty of "space" to view in the night sky. The Moon, planets, and the International Space Station all make regular appearances, but you can also see numerous stars and try to detect the patterns they make, known as constellations.

One clear night

It's amazing what you can see in just one part of the night sky. This view to the southwest, seen one March evening in Europe, has as its centerpiece the magnificent constellation of Orion. Above Orion and to its right are the planet Jupiter—one of several planets visible over the year—and two glittering star clusters, the Hyades and Pleiades. Not far from Orion is Sirius, the brightest star in the sky. With binoculars, once you know where to look you can see more types of objects, and even galaxies beyond our own.

Bright stars Castor and Pollux

Part of the constellation of Hydra

Bright star Procyon

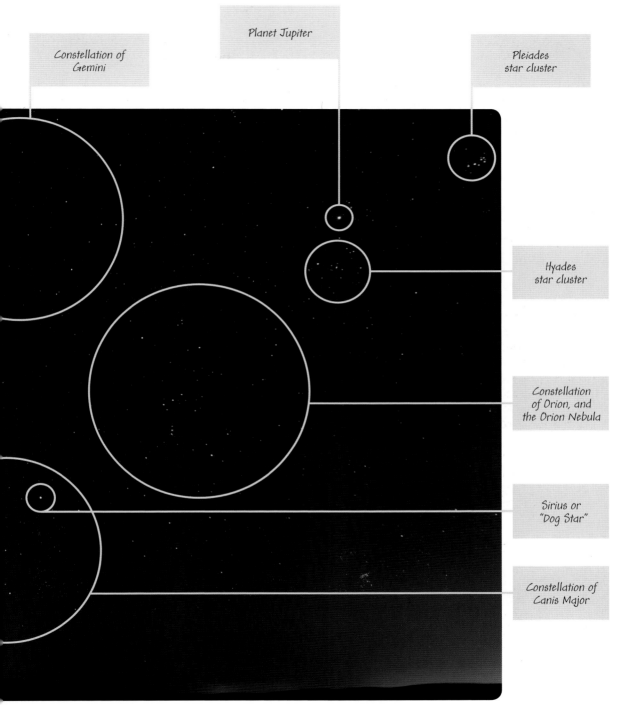

Constellation of
Gemini

Planet Jupiter

Pleiades
star cluster

Hyades
star cluster

Constellation
of Orion, and
the Orion Nebula

Sirius or
"Dog Star"

Constellation of
Canis Major

Occasional Treats

Some astronomical phenomena, such as bright nebulae (glowing clouds of dust), can be seen on any clear night—provided you know where to look. But other events are much rarer, and you may have to wait for months or years to witness them. Some phenomena, such as eclipses, are predictable years or decades in advance, but the most spectacular comets tend to appear "out of the blue."

Diary date? Perhaps the ultimate rare astronomical event is a transit of Venus, when the planet Venus is seen to pass across the face of the Sun. The last transit happened in 2012; the next is due to occur in 2117.

Showers of shooting stars

Meteors, or shooting stars, flash across the sky every night. But at various points in the year our planet encounters streams of dust that have come from bodies such as comets—and when this happens, meteor "showers" may be seen. Witnessing a shower is one of the easiest things you can do in astronomy: just lie back and look at the sky (see pp.74–75).

Total solar eclipse

A total eclipse of the Sun is rarely seen from any fixed point on Earth: to see one, prepare to travel! Lunar and partial solar eclipses are more common (see pp.32 and 90).

Bright comets

Unpredictably and infrequently—about once every 5–10 years—a dazzling comet appears in the sky for a few weeks or even months. Try not to miss such an event (see p.140).

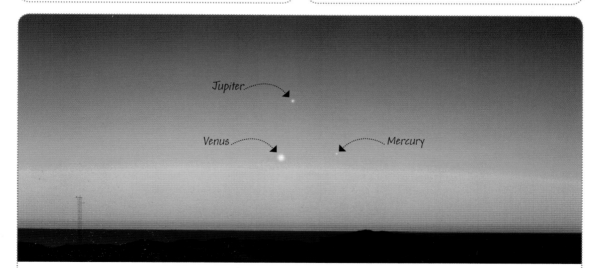

Planetary conjunctions

Every now and again an impressive spectacle occurs when several of the bright planets—Venus, Jupiter, Mars, Mercury, and Saturn—appear close together in the sky and "dance" around each other over several nights. The dates of planetary conjunctions are worth seeking out and marking in your diary. Here, Jupiter, Mercury, and Venus conjoin at sunset.

What Can You See **Through a Telescope?**

A telescope can reveal a new level of detail in the night sky, but do be realistic about what you will see. With a small telescope some objects may still only be visible as pinpoints of light or tiny fuzzy disks, and even a large one is unlikely to reveal spectacular colors.

Planetary moons

You don't need a telescope to see our Moon, but it will reveal a lot more surface detail. A small telescope may also allow you to view four of Jupiter's moons (see p.152), and some of Saturn's. The two images of Jupiter's moons shown above demonstrate how much more powerful the Hubble Space Telescope is compared with amateur equipment.

Faraway planets

Occasionally Uranus (above, left) is just visible to the naked eye as a small pinpoint of light—but through a moderate-sized telescope you may be able to make out a small blue disk.

Still more remote is Neptune (above right); with a telescope you will probably only see a tiny blue dot. A large, powerful instrument is needed to make out any detail, let alone its moons.

Distant galaxies

A few galaxies outside our own are visible to the naked eye, and more can be seen as fuzzy objects through binoculars. If you examine them with the more powerful optics offered by an amateur telescope, you may be able to make out a bit more detail. This dramatic image of the Cigar Galaxy (M82) was made using a combination of space telescopes.

What you won't see

Black holes No light ever escapes from these objects, so they are by definition invisible. Astronomers can detect black holes only by observing the effects they have on other nearby objects.

Dark matter This is thought to make up most of the matter in the Universe, but again by definition, it is invisible since it emits no light or other type of radiation. At the moment scientists are certain it exists but unclear exactly what it is.

Exoplanets These are planets orbiting stars other than the Sun. Astronomers have now found more than 900, but they can be seen —or more often detected—only with the most highly sophisticated scientific instruments.

Getting Started

The good news is that you don't need much in the way of special equipment to get started in astronomy. To avoid frustration, watch the weather forecast and pick a clear moonless night for your first outing—unless of course you're intending to observe the Moon! Seek out a good location from which to observe—for the best views, choose a spot away from light pollution with a reasonably unobstructed view of the horizon.

Make yourself comfortable

Bear in mind that you will probably be outside for long periods—often on cold nights—and looking up. It's important to wear appropriate clothing and to find a comfortable position. It takes about 20 minutes for your eyes to adapt to the dark, so you may need to be patient.

Red light
A red LED light, or a flashlight covered with a red filter or cellophane, is a useful accessory. You can see what you are doing while your eyes remain adapted to the dark.

Planning ahead
To help locate objects in the night sky, you may find it useful to acquire either a device called a planisphere or an astronomy "app" for smartphone or tablet (see pp.58–59).

Binoculars or telescope
You can see a lot with the naked eye, but eventually you may want to invest in optical aids (see pp.94 and 146).

Additional equipment
Magnetic compass to orientate yourself, unless you already have an accurate idea of the direction north or south, and can also identify other points on the compass.

Warm comfortable clothing, including a warm windproof jacket, wool hat, and fingerless gloves, unless you are observing from the tropics—in which case you may need mosquito repellent.

Deckchair for reclining comfortably as you observe; alternatively, a groundsheet and foam mattress are useful for lying flat.

Notebook and pencil for making notes and sketches of what you observe.

Thermos of coffee to help keep you warm and awake.

1

Start Simple

This chapter focuses on astronomy with the naked eye—concentrating on objects and phenomena in the night sky that you can observe without optical aids. These include the Moon, its movements and phases; the planets Jupiter and Venus; prominent stars and constellations such as Orion and the Southern Cross (Crux); meteor showers and shooting stars; and eclipses of both the Sun and Moon. You'll learn how to find your way around the night sky through "star hopping" exercises, and familiarize yourself with the concept of the celestial sphere and the use of a planisphere, which will help you understand sky movements and locate objects in the night sky.

The full Moon shortly after rising above the eastern horizon

Observing **the Moon**

Ask anyone to name a familiar object in the night sky, and the chances are they'll mention the Moon. It's visible for at least some of the time on most clear nights from anywhere in the world—and cannot be mistaken for anything else. But how well do you really know it? To kick-start your course in astronomy try keeping a lunar diary to learn about the Moon's movements.

Tracking the Moon hour by hour

Choose a moonlit night and watch the Moon over the course of a few hours. You'll see that it moves slowly toward the western horizon, where it eventually sets. This daily movement occurs because Earth is constantly rotating on its axis. Earth takes 24 hours—a day—to complete its rotation; the overall effect is that the Moon appears to move in and out of sight.

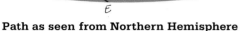

Path as seen from Northern Hemisphere   **Path as seen from Southern Hemisphere**

One Moon, two hemispheres

The Moon rises in the east and sets in the west, but in between its movement differs depending on your location. Viewed from the Northern Hemisphere it traverses the southern sky; from the Southern Hemisphere it crosses the northern sky. Features on the surface also differ: in the Southern Hemisphere (as on p.26) it's upside down to Northern Hemisphere eyes.

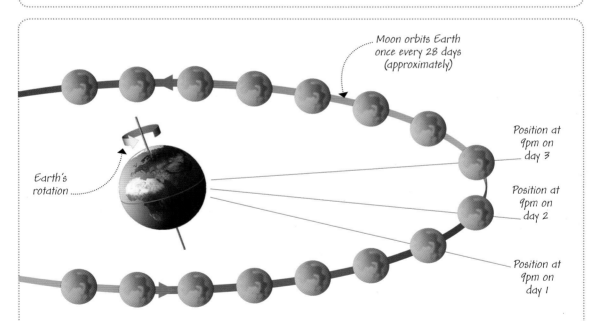

Moon orbits Earth once every 28 days (approximately)

Earth's rotation

Position at 9pm on day 3

Position at 9pm on day 2

Position at 9pm on day 1

Tracking the Moon day by day

Make a note of where the Moon is in the sky one evening at a particular time, say 9pm, and look for it again the next evening at exactly the same time. You'll find it is slightly to the east of where it was the first night—Earth's rotation has not quite brought it back to where it was 24 hours earlier. This shift to the east occurs every night and is caused by the fact that the Moon is orbiting Earth. Each 24 hours the Moon has moved around a bit in its orbit.

Phases of **the Moon**

So far you've learned why the Moon moves across the sky from east to west—because Earth spins—and also that it displays a daily west-to-east shift as it orbits Earth. It's because of this orbit that the Moon's position in the sky relative to the Sun, as seen from Earth, changes—so over about a month the area of the Moon's surface that is lit up varies from zero to fully lit. The Moon's different appearances over this cycle are described as phases. During each cycle there are also changes in when, and for how long, the Moon can be seen each night.

First half of the lunar cycle

During nights 0–14 of the lunar cycle (shown here as it appears in the Northern Hemisphere) on each successive night the sunlit area of the Moon gradually increases—it is said to be "waxing." Each night at sunset the Moon is farther to the east, and appears in the sky for about 50 more minutes before it sets.

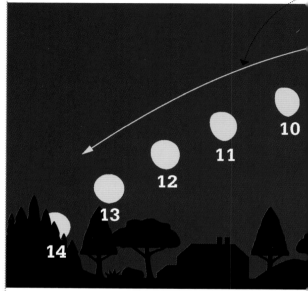

THE SKY AT SUNSET days 1–14

EAST   *Day 14: Full moon rises at sunset*

New Moon (Day 0)
In this phase the Moon lies between Earth and the Sun. The part of the Moon facing Earth is fully in shadow, and so for a few days it cannot be seen at all.

Waxing crescent (Days 1–6)
A few days later the Moon has moved to a position where a thin sliver of its surface is lit up. In this phase the Moon is visible for a few hours around and after sunset.

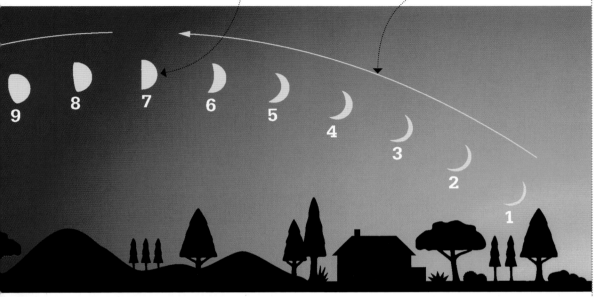

Waxing gibbous
During these phases the Moon is visible from late afternoon until the very early morning

First quarter
For Northern Hemisphere viewers, a first-quarter Moon lies south at sunset

Waxing crescent
In these phases the Moon is quite close to the Sun in the sky, slightly restricting viewing chances

9 8 7 6 5 4 3 2 1

LOOKING SOUTH   Day 0: New Moon is invisible near the Sun   **WEST**

First quarter (Day 7)

Seven days (about one quarter) into a lunar cycle, half of the Moon's face appears lit up. In this phase the Moon can be seen from the afternoon until about midnight.

Waxing gibbous (Days 8–13)

The intermediate phase between the Moon being half- and fully lit is called "gibbous." A waxing gibbous Moon can be seen from late afternoon until the early morning hours.

THE SKY AT SUNRISE
days 15–28

Waning crescent
In these phases the Moon is close to the Sun, restricting viewing opportunities

Third quarter
For Northern Hemisphere viewers, the third-quarter Moon lies south around sunrise

28   27   26   25   24   23   22   21   20

EAST   *Day 29: New Moon is invisible near the Sun*   **LOOKING SOUTH**

Full Moon (Days 14–15)
We see the Moon fully illuminated when it is on the opposite side of Earth from the Sun. In this phase it is visible all night, from sunset to sunrise.

Waning gibbous (Days 16–21)
The next phase after Full Moon is waning gibbous. In this phase the Moon rises between sunset and midnight and remains visible until it sets a few hours after sunrise.

Second half of the lunar cycle

In the second half of the lunar cycle comprising nights 15–28 (shown left, again for the Northern Hemisphere) on each successive night the area of the Moon's surface that is sunlit—as seen from Earth—gradually diminishes, and the Moon is said to be "waning." As its position in the sky shifts eastward a little every 24 hours, during this part of the cycle it becomes visible for about 50 minutes less each night because it progressively becomes blotted out by the glare of the Sun.

Southern Hemisphere observers

If you are tracking the Moon from Australia, Brazil, or other parts of the Southern Hemisphere, things will look a little bit different. Although the changes that occur over the lunar cycle are exactly the same as they are in the Northern Hemisphere, you will see the "mirror image" of what a Northern Hemisphere viewer sees. For example, in its waning phases, from the Southern Hemisphere the illuminated part of the Moon appears on the right, whereas in the Northern Hemisphere it appears on the left.

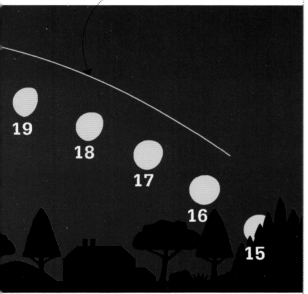

Waning gibbous
During these phases the Moon is visible from about 1–5 hours before midnight until after sunrise

Day 15: Full moon sets at sunrise   **WEST**

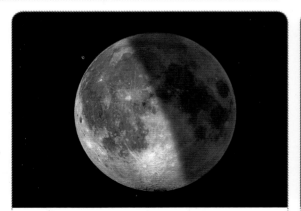

Third quarter (Day 22)

Three quarters of the way through a lunar cycle, the Moon's face is half lit up again. In this phase it rises at about midnight and is visible until noon.

Waning crescent (Days 23–28)

In the waning crescent phase, the Moon rises a few hours before sunrise. Once the Sun has also risen it tends to fade from view quite quickly due to the Sun's glare.

Observing a **Lunar Eclipse**

Once or twice a year from most parts of the world there's usually a chance to view a lunar eclipse. This can be a dramatic event, visible to the naked eye, in which Earth's full shadow covers the Moon; it occurs when the Sun, Earth, and Moon are aligned, with Earth in the middle. The table lists when and where you can see a total eclipse up to 2025.

How a lunar eclipse works

During some of its orbits around Earth, the Moon passes through Earth's shadow. Like all shadows, it varies in intensity: full shadow is known as the Earth's "umbra"; part shadow is its "penumbra." A penumbral eclipse is when the Moon passes through Earth's penumbra, resulting in a slight dimming. A partial eclipse occurs when part of the Moon passes through Earth's umbra, and during a total eclipse the whole Moon passes through the umbra.

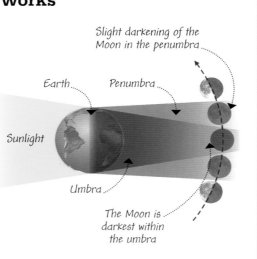

Slight darkening of the Moon in the penumbra

Earth

Penumbra

Sunlight

Umbra

The Moon is darkest within the umbra

Calendar dates: total lunar eclipses

Date	Location
April 15, 2014	Australia, Pacific, Americas
Oct 8, 2014	Asia, Australia, Pacific, Americas
April 4, 2015	Asia, Australia, Pacific, Americas
Sep 28, 2015	East Pacific, Americas, Europe, Africa, West Asia
Jan 31, 2018	Asia, Australia, Pacific, Western North America
Jul 27, 2018	South America, Europe, Africa, Asia, Australia
Jan 21, 2019	Central Pacific, Americas, Europe, Africa
May 26, 2021	East Asia, Australia, Pacific, Americas
May 16, 2022	Americas, Europe, Africa
Nov 8, 2022	Asia, Australia, Pacific, Americas
Mar 14, 2025	Pacific, Americas, western Europe, West Africa
Sep 7, 2025	Europe, Africa, Asia, Australia

Tracking a lunar eclipse

Lunar eclipses occur only when the Moon is full. The first sign is a slight darkening of the Moon as it passes into Earth's penumbra. During a penumbral eclipse—the least spectacular type—this is all that happens. In other types of eclipse, a much darker region moves across the face of the Moon as it passes into Earth's umbra (in this time-lapse photograph it can be seen happening from bottom left). During a total eclipse, which can take about an hour to develop, the Moon takes on a bloodred appearance once all of it has passed into Earth's umbra (center right in photograph). The reddening occurs because some sunlight is refracted toward the Moon by Earth's atmosphere. Subsequently the Moon slowly moves out of Earth's shadow.

Note About 2,350 years ago—when the world was believed to be flat—the Greek philosopher Aristotle put forward several arguments for believing that Earth was actually a sphere. One of his reasons was that the shadow cast upon the Moon in a lunar eclipse has a curved edge.

Understanding **How Stars Move**

So far you've tracked the Moon and learned how its phases and position
in the sky relate to its journey around Earth, our planet's own rotation, and
Earth's orbit around the Sun. Now it's time to investigate other objects in the
night sky, such as stars and constellations; but first you need to understand
a little more about how and why stars seem to move across the sky.

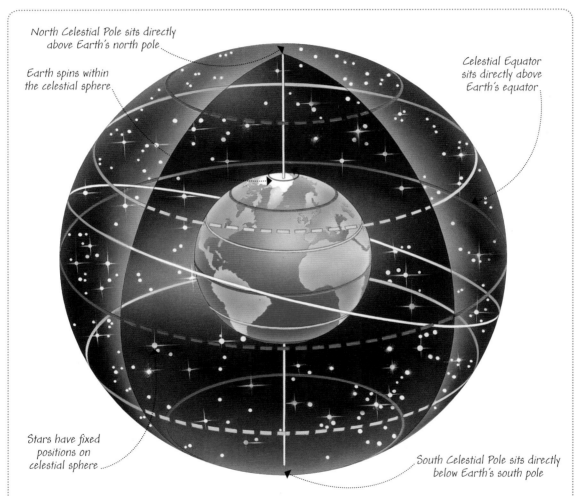

North Celestial Pole sits directly above Earth's north pole

Earth spins within the celestial sphere

Celestial Equator sits directly above Earth's equator

Stars have fixed positions on celestial sphere

South Celestial Pole sits directly below Earth's south pole

The celestial sphere

Imagine a static shell-like sphere surrounding
Earth in which all the distant stars are
embedded at fixed positions on the inside
shell. This imaginary globe is known as
the "celestial sphere"; many of its features
are "celestial" equivalents of those on Earth,
including celestial north and south poles,
and an equator (the celestial equator).

Apparent star movement

Earth is spinning continuously within the celestial sphere. This means that when you look up at the night sky you see different patterns of star movement depending on your location—whether at the north pole, the south pole, the equator, or somewhere in between.

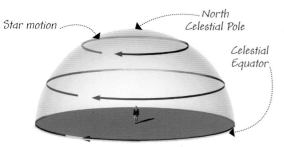

Looking up from the north pole

Stars appear to circle counterclockwise around a point directly overhead—the north celestial pole. Stars at the horizon appear to travel around the horizon.

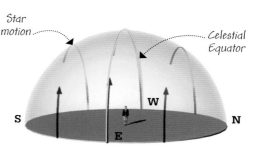

Looking up from the equator

Stars appear to rise vertically in the east, swing overhead, and then drop vertically down again and set in the west.

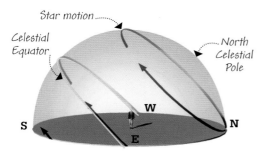

At midlatitudes (Northern Hemisphere)

Most stars rise in the east, cross the sky obliquely, and set in the west. But stars in the northernmost part of the sky move counterclockwise around the north celestial pole.

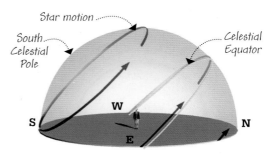

At midlatitudes (Southern Hemisphere)

Most stars rise in the east, cross the sky obliquely, and set in the west. But stars in the southernmost sky move clockwise around the south celestial pole.

Mapping the sky

The celestial sphere can be divided into six parts—two circular sections at the top and bottom centered on the celestial poles, and four sections around the middle, with the celestial equator running horizontally through. Because stars and other distant objects have a fixed place on the celestial sphere, it is possible to produce maps of these sections showing their positions. You'll find two polar sky maps and four equatorial maps on pp.176–187.

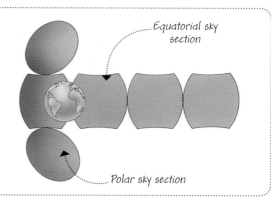

Equatorial sky section

Polar sky section

Constellations and Asterisms

Since about 1930, astronomers have divided the celestial sphere into 88 areas, called constellations. An example of a constellation—Aquila, the Eagle—is shown below. It has an irregular border, and surrounding it are other constellations, such as Delphinus and Scutum.

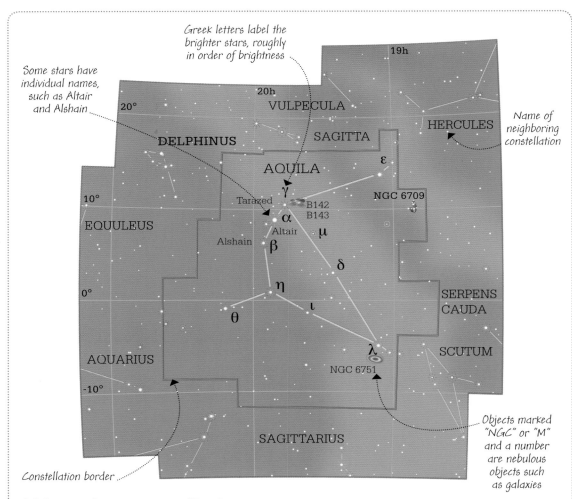

Greek letters label the brighter stars, roughly in order of brightness

Some stars have individual names, such as Altair and Alshain

Name of neighboring constellation

Objects marked "NGC" or "M" and a number are nebulous objects such as galaxies

Constellation border

Objects in a constellation

On a constellation map, the brighter stars are joined by lines to form what is called the constellation figure. Some individual stars are labeled. The oldest systematic method for naming stars uses Greek letters, with the brightest star in a constellation designated α, the second brightest with β, and so on. Other objects, such as galaxies, are also labeled.

Constellation figures

Every constellation has a figure—usually an animal (such as Aquila, the Eagle), an object, or a mythological character—and many of them represent shapes seen among groups of stars. Often they originated with the ancient Greeks, predating the regions defined by recent astronomers. They are essentially imaginary: the lines joining the stars in a figure are not seen in the sky. You'll learn how to identify some of the better known constellations as well as the more prominent stars in those figures and other objects (galaxies and nebulae) within the constellation regions.

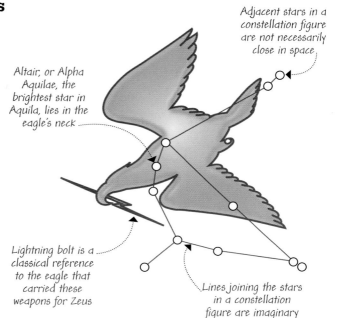

Adjacent stars in a constellation figure are not necessarily close in space

Altair, or Alpha Aquilae, the brightest star in Aquila, lies in the eagle's neck

Lightning bolt is a classical reference to the eagle that carried these weapons for Zeus

Lines joining the stars in a constellation figure are imaginary

Asterisms

Within some constellations are distinctive groupings of stars called asterisms. Examples include Orion's Belt in the constellation of Orion (see pp.60–61), the Teapot asterism in Sagittarius, the Sickle asterism in Leo, and the Big Dipper asterism in Ursa Major. You'll come across a number of these asterisms as you go through the book—they can be useful signposts helping you to find the rest of the constellation figure or other nearby sky objects.

Teapot asterism

Sickle asterism

The view **from Earth**

Earlier you learned how your position on Earth affects the pattern of star movement across the sky (see p.35). Your observing location—particularly your latitude—also affects what parts of the celestial sphere you are able to see.

Effects of latitude

Your latitude indicates how far north or south of the equator you are observing from. The diagrams below show what you can see from three different positions: at the north or south pole, at the equator, and at a point about halfway between (midlatitudes).

> **Remember** Your latitude affects the area of the sky that you can observe, but longitude makes no difference. So from New York City (US), Madrid (Spain), and Beijing (China) —all at roughly the same latitude but very different longitudes—you'll observe more or less exactly the same areas of the night sky.

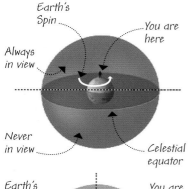

Pole positions
When you are at Earth's north pole you can only see the northern half of the celestial sphere—and it is always in view. The southern half of the sphere is never in view. If you're at Earth's south pole, then the reverse is true.

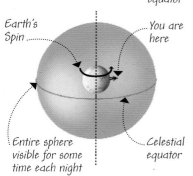

Equatorial positions
Standing at the equator, from an astronomical viewpoint you are in a much happier position: Earth's spin brings all parts of the celestial sphere into view for some of the time each night.

Positions at midlatitudes
For midlatitude locations, part of the celestial sphere is visible every night; part is sometimes in view; and a third part—a significant portion of the polar sky in the other hemisphere— will never be in view.

Northern high latitude/polar region

60°N

40°N

20°N

0°

20°S

40°S

60°S

Constellation locators

Throughout the book you'll see "locators"—diagrams like those shown here. These are intended to help you find a constellation relative to the horizon at specific times and dates at different seasons of the year. They are standardized for observing positions at 40°N or 40°S. In most cases, if you are observing the night sky from a higher latitude than 40° (north or south), the constellation will appear correspondingly lower in the sky, and vice versa.

**April 1, 7pm
Northern Hemisphere**

**April 1, 7pm
Southern Hemisphere**

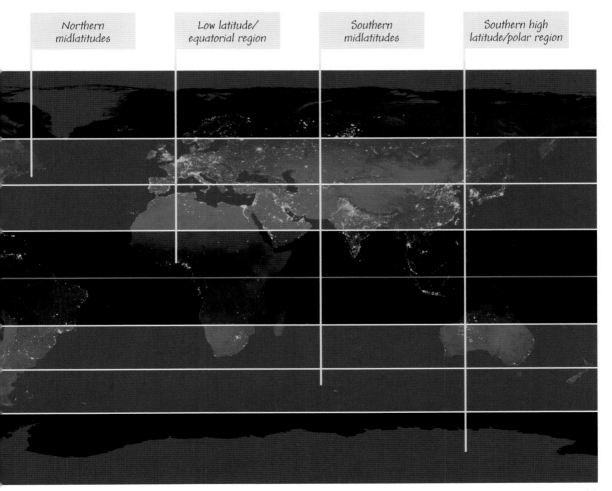

Northern midlatitudes

Low latitude/ equatorial region

Southern midlatitudes

Southern high latitude/polar region

Changes over **the Year**

You've seen how Earth's rotation brings different parts of the celestial sphere into view, but why does the night sky look different between the seasons? It's all to do with Earth's orbit of the Sun, which within a band called the zodiac (see pp.66–67) obstructs the direct view of various parts of the sky in turn.

Shifting views

The part of the celestial sphere most easily viewed at any time of year is the region on the opposite side of Earth from the Sun. Month by month, this gradually shifts. Also, in working out when to observe a particular part of the sky, there is a trade-off between date and clock time. For example, from a given location your view of the sky will be the same at midnight on November 1; at 10pm on December 1; and at 8pm on January 1.

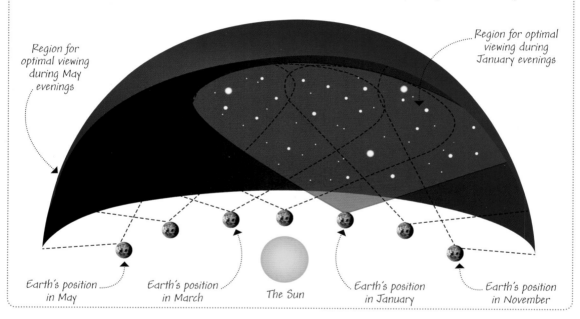

Region for optimal viewing during May evenings

Region for optimal viewing during January evenings

Earth's position in May

Earth's position in March

The Sun

Earth's position in January

Earth's position in November

Help is at hand!

If you're feeling confused about how your latitude, the time of year, and other factors such as clock time affect what you can see in the night sky—don't despair. A planisphere (right) is an astronomy aid that can help sort out these matters for you. You can learn how to use one—and find advice on some other types of astronomy aids—on pp.58–59.

Seasonal comparison

The two diagrams below compare what can be seen in the sky from Northern Hemisphere locations, looking south, in mid-March at 10pm and six months later in mid-September at 10pm. During the period between March and September, Earth travels halfway around its orbit of the Sun. This means that the view looking out at the celestial sphere completely changes, with different stars and constellations becoming prominent.

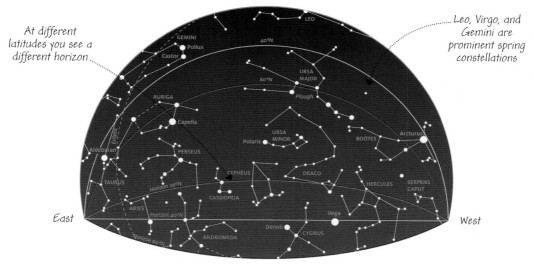

At different latitudes you see a different horizon

Leo, Virgo, and Gemini are prominent spring constellations

Looking south in mid-March at 10pm

At different latitudes you see a different horizon

Pegasus, Cygnus, Aquila, and Aquarius are prominent constellations during fall

Looking south in mid-September at 10pm

Navigating **the Night Sky**

The two points on the celestial sphere around which everything in the sky seems to continuously "wheel" due to Earth's rotation (see pp.34–35) are known as the north and south celestial poles. If you can find the appropriate pole for your hemisphere you will find it easier to navigate the night sky.

Locating the North Celestial Pole

Conveniently a star called Polaris—part of a constellation called Ursa Minor (little bear)—is very close to the north celestial pole; at any specific location in the Northern Hemisphere, Polaris is always in the same place in the sky. Even better, two stars from a nearby asterism (star pattern) called the Big Dipper point the way to Polaris.

> **Tip** If you are observing from a low latitude—areas quite far south, near the equator —you may find it easier to locate the Big Dipper at times of year other than fall.

Tracking Earth's rotation

If you observe Polaris, the Big Dipper, and Ursa Minor over several hours one evening, you should just be able to see that these and all other stars in the sky are moving slowly in a counterclockwise direction along circular paths with Polaris at the center point. Just as it was with the Moon, this apparent movement of the stars is actually caused by Earth's rotation.

Finding the Big Dipper

To locate the Big Dipper on a clear fall evening after dark, use a compass to find north, and look into the northern sky for an arrangement of stars that looks a bit like a saucepan. It should appear roughly as shown: the exact configuration depends on the time, and its height above the horizon depends on your latitude. Once you've found it, use its "pointers," the stars Merak and Dubhe, to locate Polaris and Ursa Minor. The guide diagrams below help you find the Big Dipper on winter, spring, or summer evenings: the procedure is the same, but the position of the Big Dipper (and Ursa Minor) relative to the horizon will be different. A planisphere or tablet/smartphone "app" (see pp.58–59) may help.

Winter evenings
Dec 1 at midnight; also
Jan 1, 10pm; Feb 1, 8pm

Spring evenings
March 1 at midnight; also
April 1, 10pm; May 1, 8pm

Summer evenings
June 1 at midnight; also
July 1, 10pm; Aug 1, 8pm

Locating **Crux and the South Celestial Pole**

Unfortunately there is no bright star at the south celestial pole equivalent to Polaris in the Northern Hemisphere. However a spectacular constellation called Crux, or the Southern Cross, lies near to the south celestial pole. Crux is a stunning sight in itself, but also serves as a useful "pointer" to find other interesting objects in the southern night sky (see pp.48–49 and 120–121).

Identifying Crux

Crux is a small constellation containing four bright stars in the shape of a cross. Three of the stars are appreciably brighter than the fourth. It can, however, be confused with a nearby pattern of four stars called the False Cross asterism. Crux is more compact than the False Cross, and also contains a fifth, moderately bright star within its outline.

Crux

False Cross asterism

Tracking the stars

Once you've found Crux and identified the approximate location of the south celestial pole, try observing the whole southern night sky over an hour or so. You should be able to see that all the stars move slowly in a clockwise direction along circular paths, with the center point of all the circles being the south celestial pole. If you were to take an extended-exposure photograph, the result would look something like that shown here. As with the Moon and Polaris, the apparent movement of the stars is actually due to Earth's rotation within the celestial sphere.

Finding Crux

To locate Crux on a clear early winter evening after dark, pick a good observing location and use a compass to find south. Look into the southern sky for the arrangement of four bright stars in the shape of a cross. It should appear more or less as shown, but the exact configuration will depend on the time, date, and latitude from which you're observing. The two stars at either end of the long axis of the cross, Gacrux and Acrux, point toward the south celestial pole. To get there, follow a line through the long axis of the Cross for about three "cross lengths" until you reach the relatively empty region of the night sky containing the south celestial pole. This is the point around which everything in the southern sky seems to "wheel around" (see opposite).

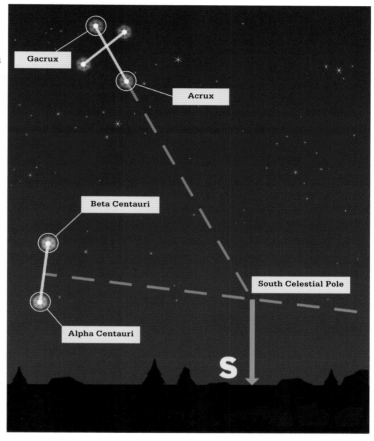

Gacrux

Acrux

Beta Centauri

Alpha Centauri

South Celestial Pole

S

Spring evenings
Aug 1 at 11pm; also
Sep 1, 9pm; Oct 1, 7pm

Summer evenings
Nov 1 at 2am; also
Dec 1, midnight; Jan 1, 10pm

Autumn evenings
Feb 1 at 11pm; also
Mar 1, 9pm; April 1, 7pm

Star hopping from
the Big Dipper

The Big Dipper (the star pattern in Ursa Major that you identified on p.42) provides a signpost to several stars and constellations. Using a star pattern to navigate to neighboring objects is called "star hopping," and it's a superb way to get familiar with the night sky. On a clear night, try these "star hops" from the Big Dipper:

1 To Polaris in Ursa Minor First, as a reminder, find Polaris, the north polar star. Visualize a line between the last two stars in the Big Dipper's "bowl," Merak and Dubhe. Extending the line past Dubhe brings you to Polaris.

2 To Arcturus in Boötes, the Herdsman Now visualize a curved line connecting the last three stars in the Big Dipper's "handle," and extend this curve until you reach an exceptionally bright star with a golden hue. This is Arcturus, the leading star in Boötes. Once you've found Arcturus, you should be able to make out the main part of the constellation figure of Boötes.

3 To Regulus in Leo Starting at Megrez, the faintest of the seven stars in the Big Dipper, draw an imaginary line to Phecda at the base of the Big Dipper's "bowl." Continuing this line, you will pass the constellation of Leo Minor, before reaching the bright white star Regulus, in Leo. Once you've found Regulus, you should be able to see the constellation figure of Leo—a crouching lion.

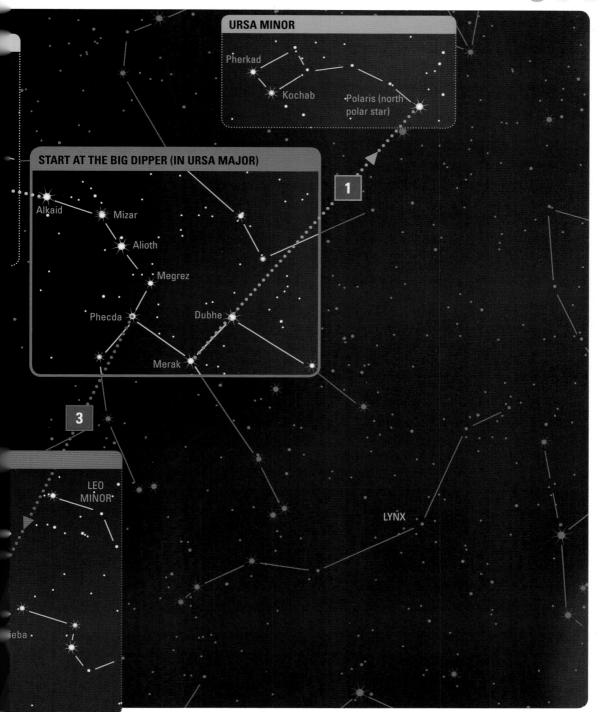

URSA MINOR

Pherkad

Kochab

Polaris (north polar star)

START AT THE BIG DIPPER (IN URSA MAJOR)

Alkaid

Mizar

Alioth

Megrez

Phecda

Dubhe

Merak

1

3

LEO MINOR

LYNX

...eba

CENTAURUS

Al Birdhaun

START AT CRUX

Gacrux

Delta Crucis

Beta Crucis

Rigil Kentaurus

Acrux

1

Hadar

2

TRIANGULUM AUSTRALE

ARA

CARINA

Alpha Trianguli Australis

HYDRA

VELA

FALSE CROSS

Kappa Velorum

Iota Carinae

Koo She

Foramen

Star hopping from
Crux

Crux is the Southern Hemisphere's equivalent of the Big Dipper—an excellent signpost for finding not just the sky's south pole, but also some neighboring constellations and asterisms. Try these "star hops":

1 **To Centaurus** Start at Delta Crucis, the fainter of the two stars on the beam of Crux. Visualize a line through Beta Crucis, the other star on the beam. You'll soon reach the bright star Hadar. Just past this is an even brighter star called Rigil Kentaurus or Alpha Centauri. These are the two leading stars in the constellation of Centaurus.

2 **To Triangulum Australe** Use the chart to find a star in Centaurus called Al Birdhaun, or Epsilon Centauri. Extend an imaginary line from this through Rigil Kentaurus; the next bright star you come to is Alpha Trianguli Australis, or Atria. This forms one vertex of a prominent triangular constellation, Triangulum Australe. The other two vertices sit equally spaced in the direction back toward Centaurus.

3 **To the False Cross asterism** Imagine a line running from Hadar in Centaurus through Acrux—the brightest star in Crux—and extend it over the Milky Way. At about twice the distance between Hadar and Acrux, you'll reach one of the stars on the False Cross's "beam," Iota Carinae. You should then be able to spot the other three stars of the False Cross.

A Quick Guide to **Stars**

Stars form from the condensation of clumps of gas and dust in galaxies: the stars that we see are part of our local galaxy, the Milky Way. Each star is an extremely hot ball of plasma (ionized gas) that generates energy through nuclear fusion reactions in its core (center). Stars often occur in clusters or as orbiting groups.

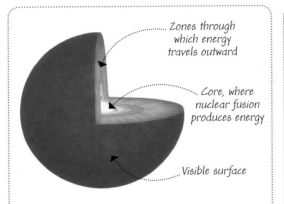

Zones through which energy travels outward

Core, where nuclear fusion produces energy

Visible surface

Structure

A star consists of an energy-producing core—which in some stars has many layers—and surrounding zones through which the energy slowly moves outward.

Stellar variety

All the stars in the night sky appear to us as tiny pinpricks of light, but in fact stars vary enormously in their size, color, and brightness, and also in their likely lifespans.

Star size and color

When stars are born, the mass of gas that forms them varies enormously. The star's mass defines not just its size but also its temperature—bigger stars are hotter—and lifespan—small stars tend to live longer. The temperature of a star, in turn, governs its color: the hottest stars are blue-white, and the coolest are red. All this means that stars come in a wide variety of types, a few of which are shown here in comparison with the Sun.

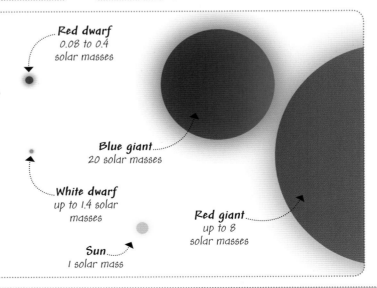

Red dwarf
0.08 to 0.4 solar masses

Blue giant
20 solar masses

White dwarf
up to 1.4 solar masses

Red giant
up to 8 solar masses

Sun
1 solar mass

Measuring magnitude (brightness)

Astronomers use two separate scales to express the brightness of stars and other sky objects. Apparent magnitude is a measure of how bright an object looks from Earth, while absolute magnitude expresses its true brightness, as if viewed from a standard distance. Amateur astronomers are most interested in apparent magnitude: faint objects are given high values on the scale, while bright objects have low or negative values.

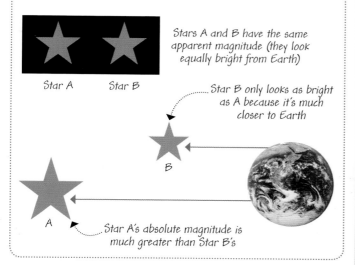

Star A   Star B

Stars A and B have the same apparent magnitude (they look equally bright from Earth)

Star B only looks as bright as A because it's much closer to Earth

B

A

Star A's absolute magnitude is much greater than Star B's

Examples of apparent magnitudes

Object	Maximum apparent magnitudes
The Sun	-26.7
The Moon	-12.9
Venus	-4.6
Jupiter	-2.9
Sirius (brightest nighttime star)	-1.4
Canopus (second brightest nighttime star)	-0.72
Alpha Centauri (star)	-0.27
Vega (star)	0.03
Rigel (star)	0.12
Polaris (star)	1.98
Ganymede (moon of Jupiter)	4.4

Rigel (in Orion)

How **Star Systems Work**

Our Sun is a solitary star, but the majority of stars in our galaxy have nearby companion stars and form binary systems of two stars, or multiple systems of three or more stars. Over time, some stars fluctuate in brightness—these are known as variable stars.

Binary and multiple stars

In binary and multiple star systems, all the stars orbit a common center of mass (or center of gravity). Various types of orbital pattern may occur, some of which are shown below. These systems of gravitationally bound stars should not be confused with pairs of stars—known as optical doubles—that appear to be close in the sky but are actually widely separated in space.

Center of mass

Equal mass
In binary systems where the two stars are of equal mass, the common center of mass lies midway between the two stars.

Unequal mass
If one star in a binary system is more massive than the other, the center of mass lies closer to the higher-mass star.

Double binary
In a double binary star system, each star orbits a companion, and the two pairs also orbit a single center of mass.

Eight-star system

Embedded in the Orion Nebula (see pp.114–115) are eight stars that are collectively known as the Trapezium Cluster, because the four brightest—visible here with pink haloes—form a trapezoid shape. Some of these stars are thought to be gravitationally bound and several are themselves binary stars, overall forming a complex system.

Variable stars

There are several reasons why the brightness of stars may vary. Occasional dips occur as one star in a binary system moves in front of the other (see Eclipsing binaries, below). Some stars have surface eruptions that make them brighter. Others, called pulsating variables, repeatedly puff up and contract, causing them to cyclically change in brightness.

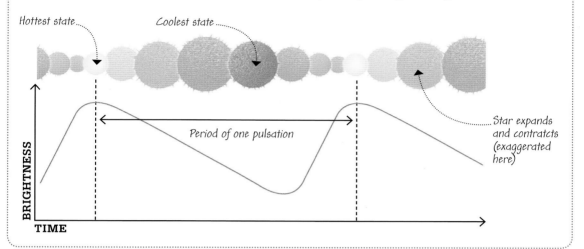

Hottest state

Coolest state

Period of one pulsation

Star expands and contratcts (exaggerated here)

BRIGHTNESS

TIME

Eclipsing binaries

In binary star systems, the overall brightness of the pair of stars may intermittently dip whenever one star moves in front of another. From the viewpoint of an observer on Earth, the star in the background is eclipsed (some light from it is partially cut off) so there is some dimming. Typically, there is a slight dimming when the fainter star in a pair is eclipsed and a more significant dimming when the brighter star is eclipsed.

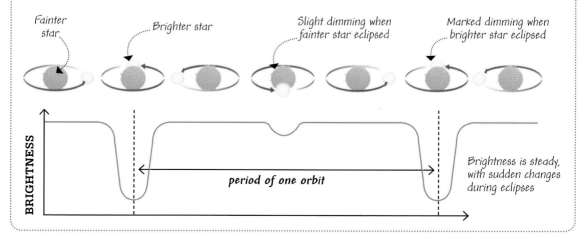

Fainter star

Brighter star

Slight dimming when fainter star eclipsed

Marked dimming when brighter star eclipsed

BRIGHTNESS

period of one orbit

Brightness is steady, with sudden changes during eclipses

Observing **the Milky Way**

On a clear night, if there is no Moon, and you can get well away from the light pollution created by towns and cities, there is a good chance you will be able to observe the Milky Way—the combined light of the vast majority of stars that make up the galaxy we live in. It is known as the Milky Way Galaxy because of the way it appears to us in the night sky.

An arc of light across the sky

The Milky Way is a swathe of faint, milky-looking light that arcs right across the sky, from one horizon to the other. It's easily mistaken for a long cloud, but unlike a cloud it does not move relative to the individual bright stars you see in the sky. Depending on your latitude, and the time and date, the band of light may or may not pass directly overhead—for example, at some times or places it may arc across the northern or southern sky.

Mapping the Milky Way

The Milky Way occupies a fixed area of the celestial sphere (see p.34): on maps it's shown as an irregular light-colored band (see below and pp.176–187). Constellation figures are also mapped; some, such as Crux and Cassiopeia, are located along the Milky Way, while others—like Ursa Minor and Boötes—are not. This can help locate and identify constellations.

Northern Hemisphere   **Southern Hemisphere**

Remember All the individual bright stars you can see in the night sky with the naked eye are part of the Milky Way Galaxy—even though when you look at a map many of them appear to be located outside the light band of the Milky Way. The reason for this is that most individual stars we can see are relatively close—within our local region of the Milky Way—and these nearby stars surround us in all directions.

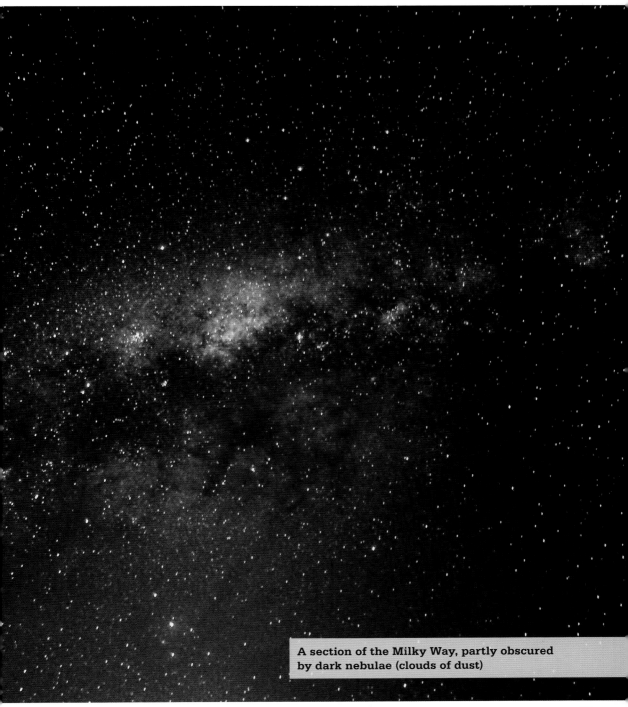

**A section of the Milky Way, partly obscured
by dark nebulae (clouds of dust)**

A Quick Guide to the Milky Way

The Milky Way is a vast whirlpool of at least 200 billion stars held together by gravity; it's around 100,000 light-years across. In shape, it resembles a pair of fried eggs held back to back, with a central bulge and a surrounding flat disk. Most of the stars are in the central bulge. Curving around this are two vast spiral arms and several smaller arms. Our Solar System lies in a minor arm known as the Orion Spur.

Galactic center

The white region in this image is Sagittarius A, a complex radio source thought to be the very center of the Milky Way. Astronomers believe that a supermassive black hole lies hidden in this region. If stars or other material are sucked into this cosmic hole by gravity, they disappear forever.

The structure of the Milky Way

We may never get a view of our galaxy from far enough away to see its overall structure. However, by studying the distribution of stars in the Milky Way, astronomers are beginning to map out its exact shape. This diagram shows our current understanding of the Milky Way.

Far 3 kiloparsec Arm

Sagittarius Dwarf Galaxy

Carina–Sagittarius Arm

Younger stars

Galactic Center
In the middle of the central hub lies a turbulent region of gas clouds and older, heavyweight stars around a supermassive black hole

Perseus Arm

Dense molecular clouds

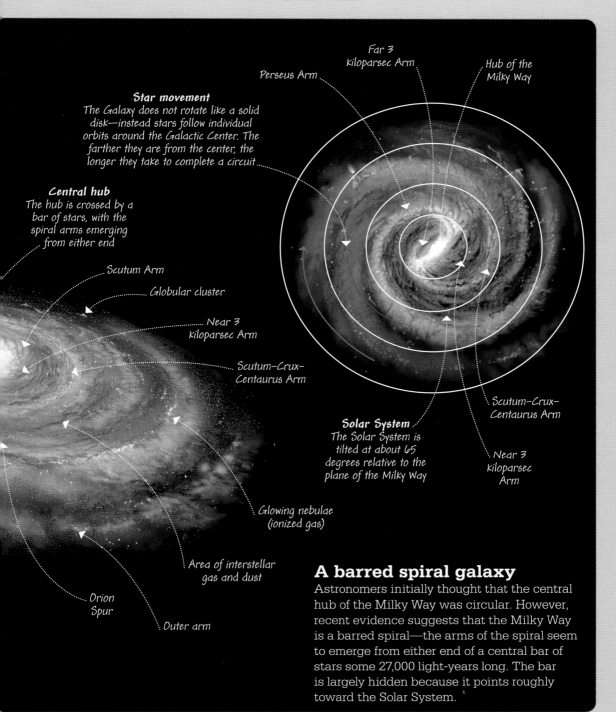

Star movement
The Galaxy does not rotate like a solid disk—instead stars follow individual orbits around the Galactic Center. The farther they are from the center, the longer they take to complete a circuit.

Perseus Arm

Far 3 kiloparsec Arm

Hub of the Milky Way

Central hub
The hub is crossed by a bar of stars, with the spiral arms emerging from either end

Scutum Arm

Globular cluster

Near 3 kiloparsec Arm

Scutum–Crux–Centaurus Arm

Scutum–Crux–Centaurus Arm

Solar System
The Solar System is tilted at about 65 degrees relative to the plane of the Milky Way

Near 3 kiloparsec Arm

Glowing nebulae (ionized gas)

Area of interstellar gas and dust

Orion Spur

Outer arm

A barred spiral galaxy

Astronomers initially thought that the central hub of the Milky Way was circular. However, recent evidence suggests that the Milky Way is a barred spiral—the arms of the spiral seem to emerge from either end of a central bar of stars some 27,000 light-years long. The bar is largely hidden because it points roughly toward the Solar System.

Locating **Stars and Constellations**

On many pages in this book you'll find locator diagrams that provide quick guidance on when, and where in the sky, you can see certain constellations, and the stars and other objects that they contain. But for more detailed guidance you may find it helpful to master using a planisphere, a smartphone or tablet astronomy "app," or other planetarium-style software.

Get your bearings

You've seen that because Earth spins on its axis and moves around the Sun our view of the stars from Earth's surface is always changing. So, the first thing to do before using astronomical aids is to check the date, the time, your latitude, and compass directions at the site from which you want to observe. Armed with this information you are ready to use a planisphere, star chart, or computer software to find out what should be visible—and where—at the time you want to observe the night sky.

Smartphone and tablet apps

Over the past 20 years or so, dedicated astronomical software has been developed, based on the principle of the planisphere. Numerous packages are available—from simple, inexpensive "apps" for smartphones and tablets that plot the positions of stars, constellations, planets, and other objects for a particular time and place, to detailed programs with vast catalogs of stars and deep-sky objects. A number of these programs allow you to switch to red night-vision mode for use outside in the dark.

Tip Some recommendations for astronomy software are given on p.191. To start with, choose a simple "app" and play with it in daylight so that you are familiar with its features and ready to use it after dark.

What is a planisphere?

A planisphere is a circular device that displays what is visible in the night sky at a particular date and time from a stated latitude. The map on the base layer is specific to that latitude, and is only suitable for observers in a band of about 5° to the north and south of it —so when choosing a planisphere it's important to check that it is appropriate for your observing location. They are available for most latitude bands except for the sparsely populated polar regions (see p.191 for recommendations).

Rotatable overlay

Base layer contains a star map

Window shows what you can see in all compass directions

Settings for date and time

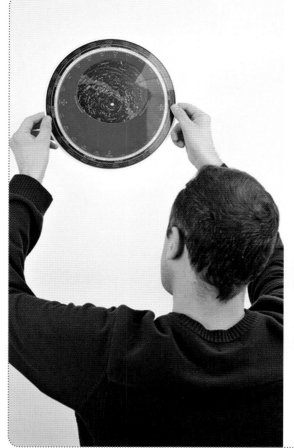

Using a planisphere

To set the planisphere, turn the overlay until the time and date are aligned at the edge. Hold the planisphere up and turn the whole device so that the horizons marked at the edge of the window match the direction in which you want to look. The constellations and stars closest to that edge are what you see nearest the relevant horizon. Planispheres can also show when to see a particular constellation or star from your observing location. Find the constellation or star on the base map, rotate the overlay until you see it in the window, ideally away from any horizon, then look at the edge for a range of dates and times.

Planisphere set for 11pm on August 12

Observing **Orion**

Orion, the Hunter, is arguably the most dramatic constellation in the sky—and a great starting point for beginners. It can be seen from both hemispheres and contains some glorious-looking objects.

Orion, the Hunter

In the classic constellation figure, Orion holds a club in one hand and a shield in the other. A glittering trio of stars forms Orion's Belt. Around the belt are four more bright stars forming the figure's distinctive hourglass shape, including the red supergiant star Betelgeuse, and the blue-white supergiant Rigel. Below Orion's Belt is the "sword" region, which contains a nebula that looks spectacular through binoculars (see pp.114–115).

Remember Orion is most easily observed from November to March. It is difficult or impossible to observe Orion from either hemisphere between May and July because at this time it is too close to the Sun.

Betelgeuse, at Orion's right shoulder, has a diameter about 1,000 times that of the Sun

The six stars that form Orion's shield are collectively known as Pi Orionist.

Orion's Belt is made up of Mintaka at the top, Anilam ("string of pearls") in the center, and Alnitak, lowest and to the left

Orion's sword includes the stunning Orion Nebula

Rigel, the brightest star in Orion, is about 40,000 times more luminous than the Sun

Northern Hemisphere locators

For observers at midlatitudes (around 40°N), Orion becomes visible from early August in the south-eastern sky—but only in the early morning hours. By December it is prominent high in the south in the late evening.

August 1, 5am
Looking southeast

December 1, midnight
Looking south

April 1, 7pm
Looking southwest

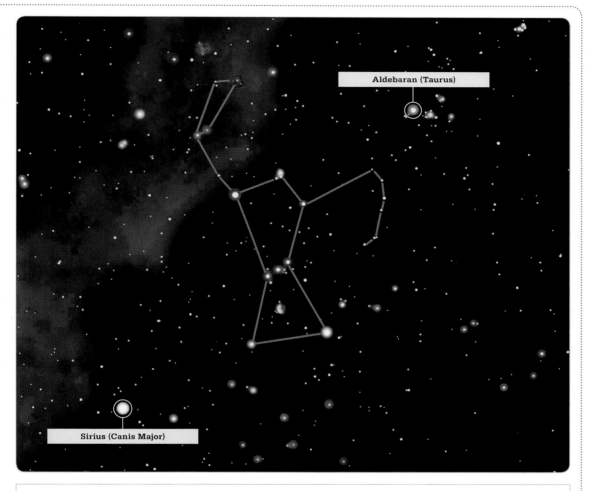

Aldebaran (Taurus)

Sirius (Canis Major)

Southern Hemisphere locators

From midlatitudes (around 40°S) Orion is visible during early mornings in August. By December it is prominent high in the north at around midnight—upside down compared to its appearance in the Northern Hemisphere.

August 1, 5am
Looking northeast

December 1, midnight
Looking north

April 1, 7pm
Looking northwest

Star hopping from **Orion**

Orion is a fantastic starting point for some "star hopping" exercises to identify nearby bright stars and the constellations in which they reside.

1 To Sirius in Canis Major Draw an imaginary line through Orion's Belt and extend it out from the same side of Orion on which Betelgeuse, the red supergiant, lies. You cannot possibly avoid stumbling upon Sirius, a dazzling pure white star that is the brightest in the entire night sky. Sirius is the leading star in the constellation of Canis Major, the Great Dog.

2 To Aldebaran in Taurus Starting again at Orion's Belt, visualize a line extending in the opposite direction, passing the star Bellatrix, which represents Orion's left shoulder. The destination is the bright red star Aldebaran—the "eye" of Taurus, the Bull. Continuing the line roughly onward, you will reach a wonderful open star cluster, the Pleiades.

3 To Procyon in Canis Minor From Bellatrix, draw an imaginary line through Betelgeuse, on Orion's other shoulder. If you follow this for just over three times the distance between Bellatrix and Betelgeuse, you should arrive at another brilliant star, yellowy-white in color, called Procyon. This is the leading star in the small and simple constellation of Canis Minor, the Little Dog. Note that Procyon, Sirius, and Betelgeuse form a large equilateral triangle.

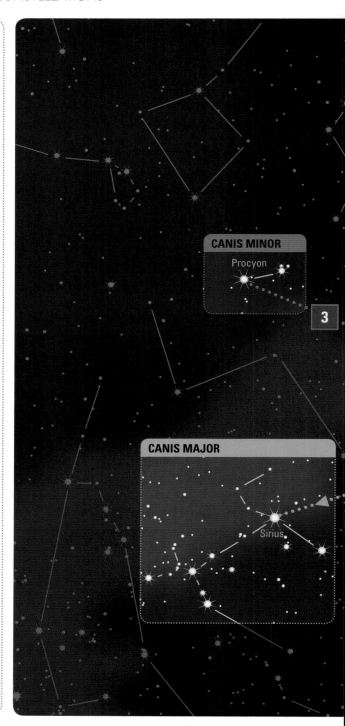

CANIS MINOR

Procyon

3

CANIS MAJOR

Sirius

PERSEUS

TAURUS

GEMINI

Pleiades

START AT ORION

Betelgeuse

Aldebaran

Bellatrix

2

Orion's Belt

1

Rigel

LEPUS

ERIDANUS

Observing **Scorpius**

The spectacular constellation of Scorpius represents the scorpion that in Greek mythology was sent by the goddess Artemis to kill Orion. It is most easily viewed from the Southern Hemisphere, but parts or even all of it may also be seen from most locations in the Northern Hemisphere.

Remember The scorpion's tail passes through an area of the Milky Way that is rich in beautiful star clusters. Although to begin with you can enjoy looking at this stunning constellation with the naked eye, it's worth returning to later with binoculars.

Scorpius, the Scorpion

Scorpius is easy to recognize. The heart of the beast is marked by the supergiant red star Antares, which is hundreds of times larger than the Sun and lies 600 light-years away. A distinctive curve of stars passing across the Milky Way represents the tail—poised ready to strike. Other bright stars represent the scorpion's head and claws.

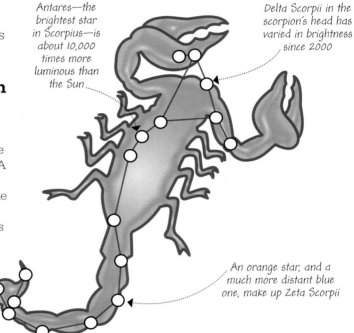

Antares—the brightest star in Scorpius—is about 10,000 times more luminous than the Sun

Delta Scorpii in the scorpion's head has varied in brightness since 2000

An orange star, and a much more distant blue one, make up Zeta Scorpii

Four moderately bright stars make up the sting in the tail of the scorpion

Northern Hemisphere locators

For observers at midlatitudes, the main possibilities for viewing Scorpius occur between April and August, looking toward the southern horizon. In June nearly all of the constellation is above the horizon around midnight.

April 1, 2am
Looking south

June 1, midnight
Looking south

August 1, 10pm
Looking south

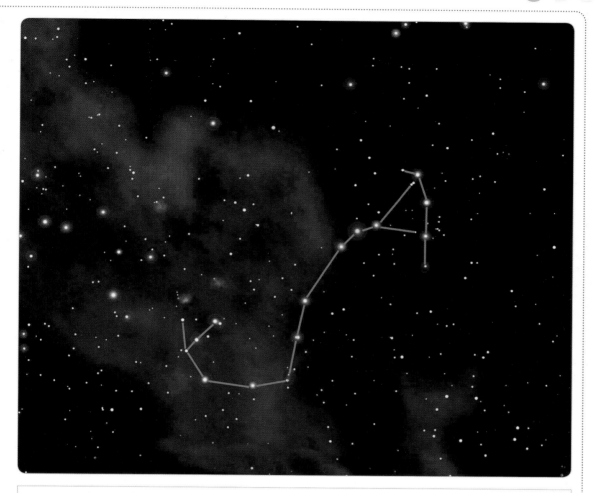

Southern Hemisphere locators

From midlatitudes Scorpius is visible from February in the early hours of the morning, looking east. By June it is high in the north or overhead in the late evening. In September it can still be seen earlier in the evening, looking west.

March 1, 2am
Looking east

June 1, midnight
Looking high in north

September 1, 10pm
Looking west

A Quick Guide to **the Zodiac**

Although it's not obvious because of the Sun's glare, as Earth orbits the Sun, the position of the Sun gradually moves day by day against the background of stars on the celestial sphere. The path traced out by the Sun is called the ecliptic, and a band of the celestial sphere on either side of this path is known as the zodiac.

Movements through the zodiac

As seen from Earth, all of the planets also gradually move, at varying speeds, against the starry background. Unlike the Sun, they do not strictly follow the line of the ecliptic, but they always stay within the band of the zodiac. Their movements are predominantly in the same direction as the Sun. In these movements, both Sun and planets pass through 13 constellations, known as zodiacal constellations.

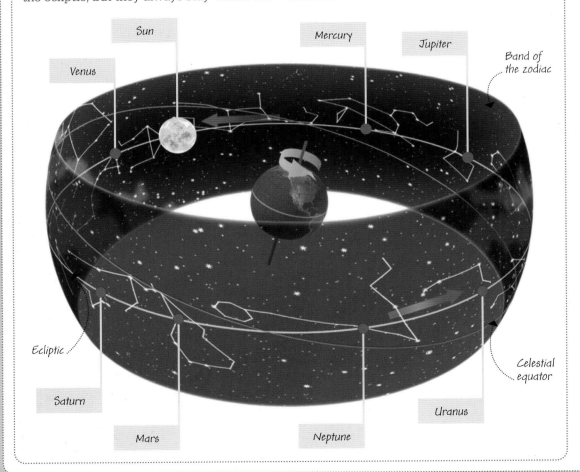

Sun

Venus

Mercury

Jupiter

Band of the zodiac

Ecliptic

Celestial equator

Saturn

Mars

Neptune

Uranus

Zodiac signs

Long ago astrologers adopted 12 of the 13 zodiacal constellations as "signs of the zodiac." For obscure reasons, the thirteenth—Ophiuchus— was largely ignored by astrologers.

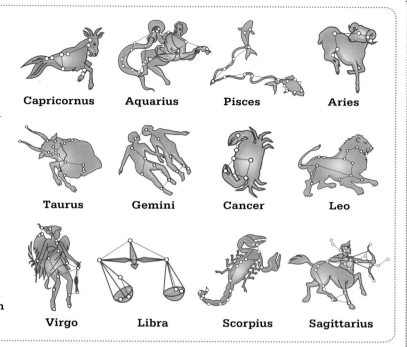

Capricornus   **Aquarius**   **Pisces**   **Aries**

Taurus   **Gemini**   **Cancer**   **Leo**

Virgo   **Libra**   **Scorpius**   **Sagittarius**

Ophiuchus, the thirteenth zodiacal constellation

Retrograde motion

The planets predominantly travel through the zodiac in the same direction as the Sun, but sometimes for a few weeks or months one appears to move in the opposite direction. This is "retrograde motion," and mostly occurs when Earth overtakes another planet "on the inside" as they orbit the Sun. For example, Earth orbits faster than Mars and "overtakes" it every 25 months. As this happens, against the background of stars Mars seems to switch its direction of travel but resumes its normal motion once Earth has swung by.

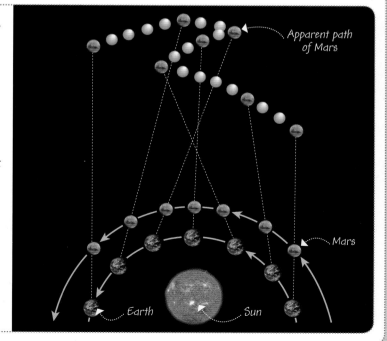

Apparent path of Mars

Mars

Earth   *Sun*

Observing **Jupiter**

Jupiter is one of the easiest planets to spot with the naked eye: when present in a clear sky it shines brighter than any star, with a steady white light. It is closest to Earth—and at its brightest—during "opposition."

Tracking Jupiter

The easiest way to spot Jupiter is to look for it in the night sky around the time of opposition. The chart below shows the months up to 2030 when the planet is at—or close to—opposition. The constellation that Jupiter is "in" at the time of opposition is also given—so in March 2016 Jupiter is situated just below the constellation figure of Leo.

To locate Jupiter around opposition, look into the southern sky (from the Northern Hemisphere) or overhead at around midnight; if you are in the Southern Hemisphere, look into the northern sky. Jupiter will be the brightest object in view, other than the Moon.

> **Remember** Jupiter only moves between constellations about once a year. When it's not in opposition, look for the nearest constellation figure and you may be able to spot Jupiter too.

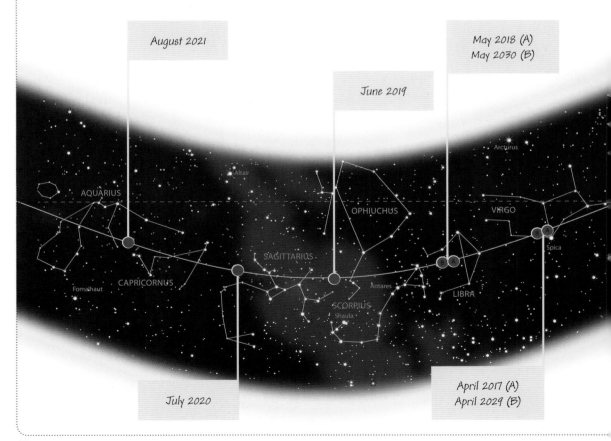

August 2021

June 2019

May 2018 (A)
May 2030 (B)

April 2017 (A)
April 2029 (B)

July 2020

The view from Earth

Jupiter takes 12 years to orbit the Sun. As Earth takes just one year, it overtakes Jupiter "on the inside" every 13 months: this is known as "opposition," and is the best time to see the planet. Jupiter cannot be seen when it is in "conjunction" on the other side of the Sun, which also happens every 13 months.

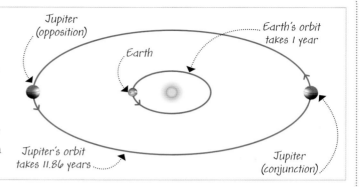

Jupiter (opposition)

Earth

Earth's orbit takes 1 year

Jupiter's orbit takes 11.86 years

Jupiter (conjunction)

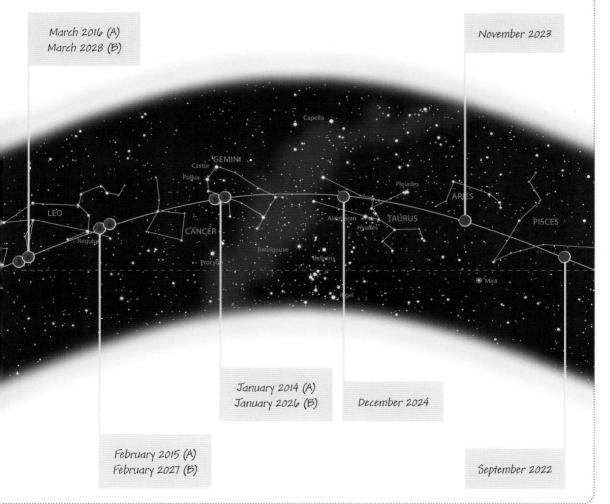

March 2016 (A)
March 2028 (B)

November 2023

January 2014 (A)
January 2026 (B)

December 2024

February 2015 (A)
February 2027 (B)

September 2022

A Quick Guide to **Jupiter**

Easily the largest planet in the Solar System, Jupiter's mass is almost 2.5 times that of the other planets combined. It has a turbulent atmosphere with winds up to 388 mph (625 kph), and the strongest magnetic field of all the planets. Thanks to immense gravity, Jupiter has amassed a huge collection of moons.

Orbit

Jupiter's orbit is highly elliptical—its distance from the Sun varies by 47.2 million miles (76.1 million km). It tilts only slightly as it spins; neither hemisphere ever points markedly toward or away from the Sun, so the planet has no seasons. Jupiter has a particularly fast spin, which causes a distinct bulge at its equator.

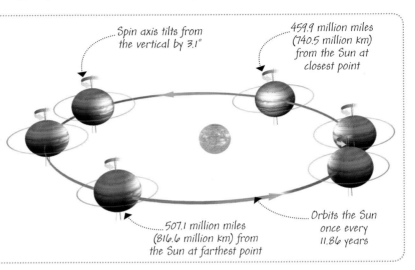

Spin axis tilts from the vertical by 3.1°

459.9 million miles (740.5 million km) from the Sun at closest point

507.1 million miles (816.6 million km) from the Sun at farthest point

Orbits the Sun once every 11.86 years

Structure

Core of rock, metal, and hydrogen compounds

Liquid metallic hydrogen

Liquid hydrogen and helium

Gaseous hydrogen and helium

Jupiter's center is a dense core thought to consist partly of rock. Surrounding this are liquid layers, mainly of hydrogen and helium, which gradually merge into an atmosphere of hydrogen and helium gas.

The Great Red Spot

The largest and most obvious single feature on Jupiter's surface is the Great Red Spot, a vast persistent storm that has been raging for at least 300 years. The whole spot rotates counterclockwise about once every six days.

Jupiter profile

Jupiter is the fifth planet out from the Sun, and with a diameter of 88,846 miles (142,984 km) it is the largest planet in the Solar System.

Average distance from the Sun
483.6 million miles (778.3 million km)

Orbits the Sun once every
11.86 years

Rotates once every
9 hours 55 minutes 30 seconds

Greatest apparent magnitude
-2.9 (see p.51)

Farthest from Earth
601.5 million miles (968.1 million km)

Nearest to Earth
365.7 million miles (588.5 million km)

Greatest angular size in sky
50.1 arcseconds (see p.12)

Number of moons
67 confirmed as of July 2013

Size comparison

Earth

Jupiter

Distinctively banded atmosphere, affected by fierce storms and winds

The Great Red Spot is about 18,600 miles (30,000 km) wide

Cloud-top temperature of -160°F (-110°C)

Observing **Perseus**

A prominent northern constellation, Perseus represents the mythological hero who slayed the gorgon Medusa. If you're in the Northern Hemisphere, this is one of the first constellations to identify and take a look at—the star Algol is a particularly interesting feature to observe because it periodically fluctuates in brightness.

Perseus, the Greek hero

The figure of Perseus is depicted with his sword in one hand and Medusa's severed head in the other. The bright star Mirphak—on the hero's abdomen—lies at the center of a cluster of fainter blue stars. Close to the sword handle is a double cluster of stars, which appear as a brighter patch in the Milky Way.

> **Remember** Once you've identified Perseus, look at the star Algol over successive nights. Every 69 hours it fades in brightness for about 10 hours. This is because it consists of two stars circling each other—it's an example of an eclipsing binary (see p.53).

Miram is an orange supergiant star with a fainter blue companion

Mirphak, a yellow-white supergiant, is the brightest star in Perseus

Algol, on Medusa's brow, is a famous variable star

Rho Persei is a red giant that fluctuates in brightness every 7 weeks

Xi Persei, a blue giant star, marks Perseus' shin

Northern Hemisphere locators

For an observer at midlatitudes (40°N), Perseus is visible at some point during the night for most of the year. The best seasons to view it are fall and winter, when it is high in the south or overhead in the late evening.

July 1, 4am
Looking east

November 1, midnight
Looking high in south

March 1, 8pm
Looking west

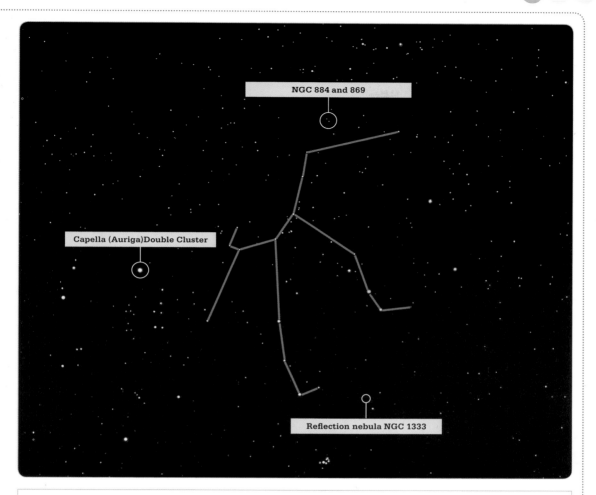

NGC 884 and 869

Capella (Auriga) Double Cluster

Reflection nebula NGC 1333

Reflection nebula NGC 1333

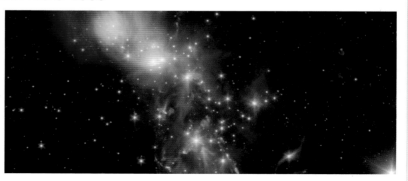

Near Perseus' left foot is a cloud of dust that reflects light and other radiation from nearby stars. This "reflection nebula" is about 1,000 light-years away. The red and green glows of infrared light were detected by NASA's Spitzer Telescope.

Observing a **Meteor Shower**

Meteors are the name astronomers give to shooting stars—small grains of space dust that hit Earth's atmosphere at enormous speed and then burn up, producing streaks of light that shoot across the sky. At certain times of year, "showers" of meteors—many flashes, all emanating from the same point in the sky, called the radiant—can be observed.

When and where to observe a shower

First identify the next meteor shower using the table below, bearing in mind the time of year and whether you're observing from the Northern or Southern Hemisphere. Choose a clear night on, or close to, the peak date—preferably one without a full or waning gibbous Moon (see p.30)—and find a spot well away from light pollution with a good view of the full sky from which to observe. Start watching for meteors around midnight or in the early hours of the morning. By then the radiant for most showers will be fairly high above the horizon.

Observation technique

First look into the sky and try to identify the constellation that contains the radiant—the point from which most of the meteors will emanate. The table below lists the relevant constellation for each shower. If necessary use a planisphere (see p.59) to help locate it in the sky. Give your eyes 20 minutes to adapt to the dark. Then sit or lie back, look up, and be patient. The flashes can happen anywhere in the sky and are usually quick and unpredictable—but over an hour or so you should see plenty.

Calendar dates: major annual meteor showers

Date range	Peak dates	Constellation	Shower	Max per hour
Dec 28–Jan 12	Jan 3–4	Boötes (N)	Quadrantids	100
Apr 16–26	Apr 22	Lyra (N)	Lyrids	20
Apr 21–May 24	May 6	Aquarius (S)	Eta Aquariids	35
Jul 14–Aug 18	Jul 28–29	Aquarius (S)	Southern Delta Aquariids	20
Jul 15–Aug 25	Aug 1	Capricornus (N, S)	Alpha Capricornids	5
Jul 23–Aug 22	Aug 12	Perseus (N)	Perseids	80

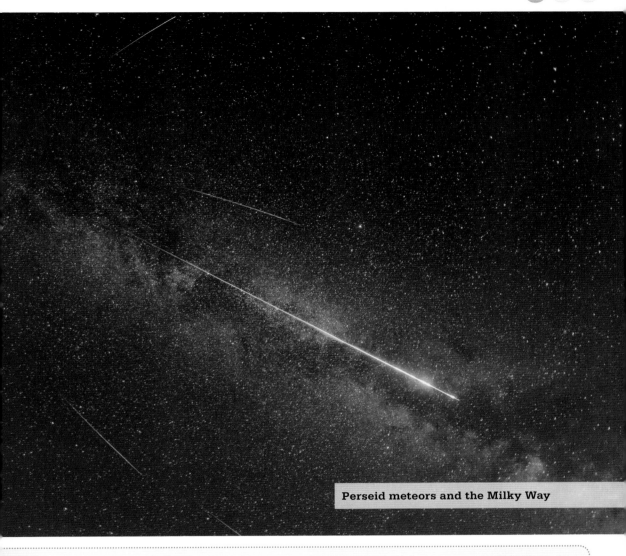

Perseid meteors and the Milky Way

Date range	Peak dates	Constellation	Shower	Max per hour
Sep 27–Dec 2	Oct 30–Nov 7	Taurus (N, S)	Taurids	10
Oct 5–30	Oct 20	Orion (N, S)	Orionids	35
Nov 14–21	Nov 17	Leo (N, S)	Leonids	Variable
Dec 6–18	Dec 14	Gemini (N)	Geminids	100
Dec 17–25	Dec 22	Ursa Minor (N)	Ursids	102

(N) Best viewed from Northern Hemisphere   (S) Best viewed from Southern Hemisphere   (N, S) Viewable from either hemisphere

A Quick Guide to **Meteors and Meteorites**

Moving around in space within the Solar System are large amounts of dust and rock that have come from comets and fragmented asteroids. When a small grain of this material hits Earth's atmosphere and burns up, it is called a meteor; but if a large chunk penetrates the atmosphere and reaches the ground without vaporizing, it is termed a meteorite.

Meteor showers

Meteors, or shooting stars, can be seen on any night, but showers tend to happen at certain times of year when Earth, in its orbit around the Sun, passes through a trail of dust left by a comet. For example, the Perseid meteor shower—which peaks in August—occurs when Earth passes into a dust trail left by comet Swift-Tuttle. The dust trail has a specific orientation in space, which means that all the shooting stars seem to originate from the same point in the sky, called the radiant.

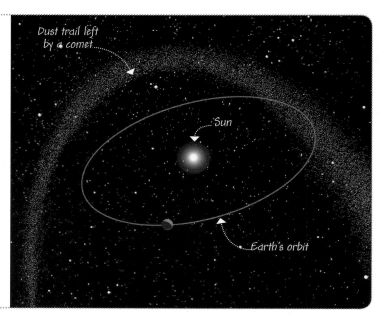

Dust trail left by a comet

Sun

Earth's orbit

Meteorites

Meteorites are classified according to their composition. Iron meteorites consist mainly of an iron-nickel alloy, stony meteorites are rocky, and stony-iron ones are a mixture of silicate and iron-nickel alloy. Extremely rarely, a meteorite injures a person when it lands. On even rarer occasions, a large meteorite may create a massive impact crater.

Iron meteorite

Stony meteorite

Stony-iron meteorite

Geminid Meteor over Iran

Observing **Cygnus**

During summer in the Northern Hemisphere, a beautiful constellation to look for is that of Cygnus, the Swan. Its brightest star, Deneb, also belongs to a magnificent trio of stars known as the Summer Triangle. If you can spot this triangle, it will help you locate Cygnus as well.

Cygnus, the Swan

The constellation figure depicts a swan flying southward along the Milky Way—on a dark night you should be able to see this in the background. Deneb sits in the swan's tail. The other two stars of the Summer Triangle are in nearby constellations: they are Vega (Lyra), and Altair in Aquila.

Kappa Cygni, a yellow giant star, lies at the tip of the swan's left wing

The North America Nebula is to the right of the swan's tail

Deneb is a blue-white supergiant

Rukh, or Delta Cygni, is a triple star system

Eta Cygni is close to an object that gives off X-rays: it may be the site of a black hole

Albireo is a double star consisting of a golden giant and a smaller blue companion

Remember Deneb may be 200,000 times more luminous than the Sun, but from Earth it looks fainter than the other two stars in the Summer Triangle. This is because it is more than 1,500 light-years away compared to 25 and 17 light-years for Vega and Altair.

Northern Hemisphere locators

From midlatitudes (around 40°N), Cygnus is visible at some time during the night all year round. In summer it is almost overhead around midnight, and from fall to mid-winter you should see it in the western sky earlier in the evening.

May 1, 2am
Looking east

August 1, midnight
Looking high in south

November 1, 10pm
Looking west

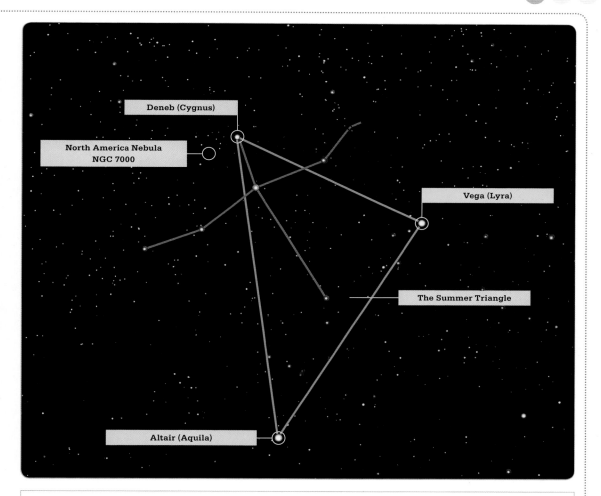

Deneb (Cygnus)

North America Nebula
NGC 7000

Vega (Lyra)

The Summer Triangle

Altair (Aquila)

North America Nebula NGC 7000

Close to Deneb is a slightly brighter patch of the Milky Way. Known as the North America Nebula after its resemblance to the continent's shape, this is an emission nebula (see p.116). You may need to use binoculars to see it clearly.

Star hopping from **the Summer Triangle**

The brightest three stars in northern summer skies—Deneb, Vega, and Altair—form a distinct triangular signpost that can be used to identify nearby constellations. Try these "star hops."

1 To Hercules From Deneb, the faintest star in the Triangle, follow an imaginary line along the shortest side of the Triangle to Vega. Continue for about the same distance again, and you will come to the constellation of Hercules, the Strong Man. Look out for a distinctive four-star asterism within Hercules—the Keystone—just off to the side of your imaginary line.

2 To Nunki in Sagittarius Starting from Deneb, trace a line through Altair, the third star in the Summer Triangle. Continuing this line about the same distance again, and deviating a little to the east, you will arrive at a star called Nunki, which is the second brightest star in the constellation of Sagittarius.

3 To the head of Draco Deneb lies on the "tail" of Cygnus, the Swan. Using the chart to help if necessary, find the central star of Cygnus, called Sadr. From here visualize a line to the star Rukh, on the wing of the Swan. Extend the line on and you will come to two bright stars in the constellation of Draco, the Dragon. Next to these are two fainter stars; together the four stars make up the head of the Dragon.

URSA MAJOR

DRACO

Thuban

Polaris

CEPHEUS

CASSIOPEIA

Observing **Venus**

Venus, our nearest neighbor, is unmistakable and spectacular when present in the evening or early morning sky. It outshines everything except the Sun and Moon.

The view from Earth

Venus orbits the Sun faster than Earth, overtaking it "on the inside" every 19 months. We don't see it at the point of each overtake (inferior conjunction) because it sits between Earth and the Sun, but afterward, for about 6 months, Venus may be seen in the east at dawn. It then moves out of view on the other side of the Sun (superior conjunction) for a few months. Venus next appears in the west around and after sunset for 6 months until it overtakes Earth again.

Locating Venus

To view Venus, choose a date within 6 weeks of "greatest elongation"—Venus will be at its maximum distance from the Sun, and easier to see. This is not necessarily when the planet is at its brightest; for evening apparitions this follows a few weeks later. You need binoculars or a telescope to see the Moon-like phases Venus displays over each 19-month cycle.

LOOK EAST in the early morning, before and around dawn

Date	Brightest	Greatest Elongation
Jan–Aug 2014	Feb 3–24	Mar 22
Sep 2015–Mar 2016	Sep 10–Oct 6	Oct 26
Apr–Oct 2017	Apr 18–May 7	Jun 3
Nov 2018–May 2019	Nov 20–Dec 14	Jan 6
Jun 2020–Jan 2021	Jun 26–Jul 28	Aug 13
Feb–Aug 2022	Jan 31–Feb 22	Mar 20
Sep 2023–Mar 2024	Sep 8–Oct 2	Oct 23
Apr–Oct 2025	Apr 15–May 5	Jun 1

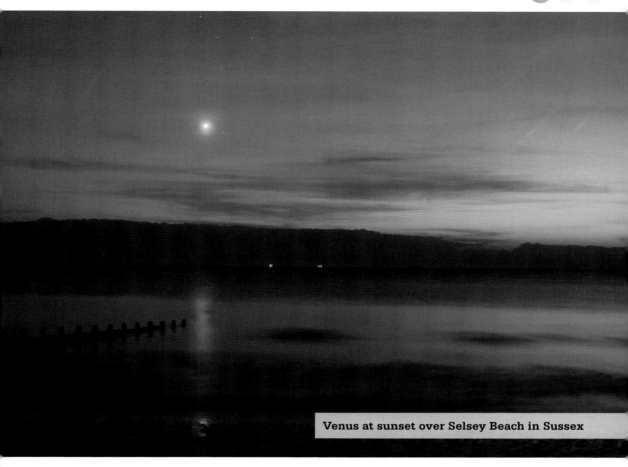

Venus at sunset over Selsey Beach in Sussex

LOOK WEST in the evening, around and after sunset

Date	Greatest Elongation	Brightest
Feb–Aug 2015	Jun 6	Jun 22–Jul 24
Sep 2016–Mar 2017	Jan 12	Feb 7–27
Apr–Oct 2018	Aug 17	Sep 9–Oct 5
Nov 2019–May 2020	Mar 24	Apr 18–May 5
Jun 2021–Jan 2022	Oct 29	Nov 22–Dec 18
Feb–Jul 2023	Jun 4	Jun 20–Jul 22
Aug 2024–Feb 2025	Jan 10	Feb 4–24
Apr–Oct 2026	Aug 15	Sep 8–Oct 3

Tip If you look into the sky around or after sunset, or around dawn as appropriate on your chosen date, you will normally have no problem identifying Venus as by far the brightest object in that part of the sky. You are likely to see it again in roughly the same place on several subsequent evenings or mornings.

A Quick Guide to **Venus**

Venus is similar to Earth in size and structure, but differs greatly in other respects. An unbroken blanket of dense clouds permanently envelopes the planet, and underneath lies a gloomy, scorching hot, waterless, and lifeless surface. The volcanic activity that shaped Venus is thought to have occurred some 500 million years ago.

Orbit

Venus has the most nearly circular orbit of all the planets. It spins in the opposite direction from the other planets (except for Uranus, which spins on its side.) The rotation period—more than 243 Earth days—is longer than that of any other planet, so a day on Venus lasts longer than the its year (around 225 Earth days).

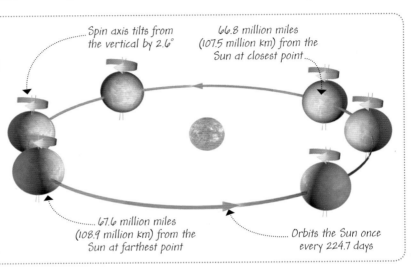

Spin axis tilts from the vertical by 2.6°

66.8 million miles (107.5 million km) from the Sun at closest point

67.6 million miles (108.9 million km) from the Sun at farthest point

Orbits the Sun once every 224.7 days

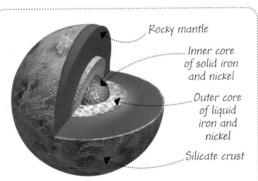

Rocky mantle

Inner core of solid iron and nickel

Outer core of liquid iron and nickel

Silicate crust

Structure

Venus has a solid metallic inner core, a molten metallic outer core, and a thick rocky mantle overlain by a thin crust. An extremely thick atmosphere—mainly carbon dioxide—surrounds the planet.

Maat Mons

Images of Venus's surface sent back by probes reveal a barren landscape shaped by huge volcanoes. Maat Mons is the largest volcano on Venus, rising 3 miles (5 km) above the surrounding plains.

Venus profile

Venus is the second planet out from the Sun. With a diameter of 7,521 miles (12,104 km) it is the third smallest planet in the Solar System.

Average distance from the Sun
67.2 million miles (108.2 million km)

Orbits the Sun once every
224.7 days

Rotates once every
243 days 16 hours 28 minutes

Greatest apparent magnitude
-4.6 (see p.51)

Farthest from Earth
162 million miles (261 million km)

Nearest to Earth
23.7 million miles (38.2 million km)

Greatest angular size in sky
66 arcseconds (see p.12)

Number of moons
0

Size comparison

Venus   Earth

Hostile, desolate surface with no liquid water

Vast ancient lava flows

Average surface temperature of 867°F (464°C)

Observing **Artificial Satellites**

Far above Earth's surface are thousands of objects that have been put there by scientists. Known as artificial satellites, they range from simple laser-reflecting spheres to complex space telescopes and a massive manned spacecraft, the International Space Station (ISS). It's easier to observe some of these objects from Earth than you might think.

Satellite orbits

Satellites follow a variety of different orbits around Earth. These orbits are established at launch and continuously fine-tuned by signals sent from control stations on Earth.

Satellites in low Earth orbit—most notably the International Space Station and Hubble Space Telescope (HST)—are most easily seen from the ground as pinpricks of light, caused by the reflection of sunlight, moving steadily across the sky. Different low Earth satellites cross the sky in various directions—from east to west, north to south, or obliquely.

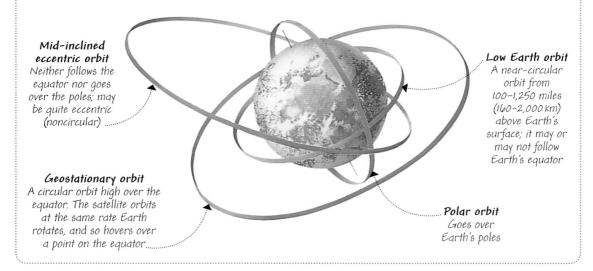

Mid-inclined eccentric orbit
Neither follows the equator nor goes over the poles; may be quite eccentric (noncircular)

Geostationary orbit
A circular orbit high over the equator. The satellite orbits at the same rate Earth rotates, and so hovers over a point on the equator.

Low Earth orbit
A near-circular orbit from 100–1,250 miles (160–2,000 km) above Earth's surface; it may or may not follow Earth's equator

Polar orbit
Goes over Earth's poles

Iridium flares

Many artificial satellites cross the sky each night, but one group is responsible for the spectacular phenomena known as "Iridium flares"—dazzling flashes of light lasting up to 10 seconds. Flares occur when the antennae of Iridium communications satellites reflect an intense beam of sunlight toward Earth; they are predictable, and internet calculators will tell you when to expect one in your area.

International Space Station

The ISS is the most noticeable of all satellites —a sharp, very bright starlike dot that moves across the sky at a slow, steady pace. Typical passes last 1–7 minutes: the ISS may traverse a large area of sky before disappearing into Earth's shadow, or over the horizon.

For a particular observing location there tend to be regular "flyby seasons" for the ISS that last for about a week, during which the space station may make several visible passes, usually soon after sunset, or before sunrise, when it is basking in sunlight. NASA offers a web-based tool listing when the ISS will be visible from your area, as well as the location in the sky where it will first appear and where it will move to before disappearing. A similar tool exists listing times to view the Hubble Space Telescope (HST).

The science of **Gravity and Orbits**

Gravity is a force that attracts objects—its strength depends on the masses of the bodies and distance between the centers of mass. The gravitational attraction between everyday-sized objects, such as billiard balls placed next to each other, is tiny. But if at least one object is big—like a planet—or the overall amount of matter is large, gravity has a profound effect.

What does gravity do?

Some important effects of gravity are shown below. Gravity holds large objects, such as stars, together, and keeps various bodies in orbit around others. Because gravity causes matter to aggregate, without it galaxies, stars, and planets would never have formed. Instead, the Universe would consist of just a thin—although vast—cloud of gas.

Holds stars together: without it, a star would disperse into space

Keeps planets in orbit around the Sun

Keeps everything in the Milky Way Galaxy in orbit around the galactic center

Keeps the Moon in orbit around Earth

Newton's idea about gravity

The scientist Isaac Newton was the first person to realize that the same force that causes an apple to fall to the ground on Earth—gravity—also causes the Moon to orbit Earth, Earth and other planets to orbit the Sun, and so on. Newton reasoned that an object remains stationary or moves at a constant speed in a straight line unless a force acts upon it. In the case of the Moon, the force of gravity continuously acts to pull the Moon toward the Earth. This pull changes what would otherwise be the Moon's straight-line motion into an orbit, as shown in the diagram (right).

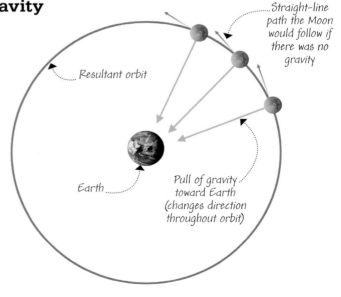

Straight-line path the Moon would follow if there was no gravity

Resultant orbit

Earth

Pull of gravity toward Earth (changes direction throughout orbit)

Common center of gravity

We usually think of a small object orbiting a much larger one, but in space there are many examples of two bodies—typically stars—of more equal size that orbit around each other.

More precisely, the two bodies each orbit the overall center of gravity, sometimes called the common center of mass, of the whole system, as shown in the diagram below.

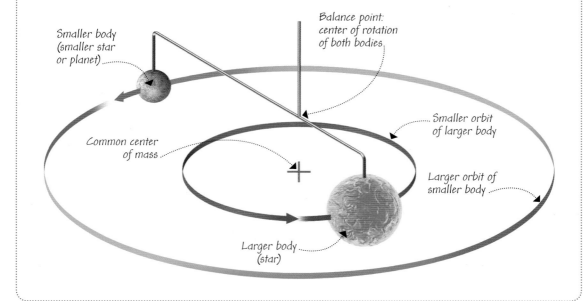

Smaller body
(smaller star
or planet)

Balance point:
center of rotation
of both bodies

Smaller orbit
of larger body

Common center
of mass

Larger orbit of
smaller body

Larger body
(star)

Orbit shapes and eccentricity

All naturally occurring orbits have the shapes of ellipses—"stretched" circles. The degree to which an elliptical orbit differs from a perfect circle is called its "eccentricity." Orbits that are quite close to being circles—such as the

Moon's orbit around Earth—are said to be of low eccentricity, while orbits that are very elongated ellipses are of high eccentricity. The orbits of comets around the Sun, for example, vary from highly to exceedingly eccentric.

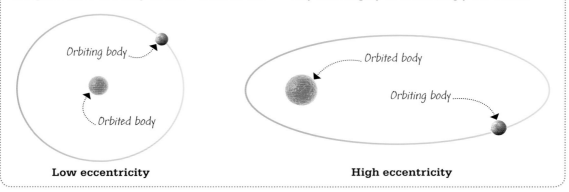

Orbiting body

Orbited body

Low eccentricity

Orbited body

Orbiting body

High eccentricity

Observing a **Solar Eclipse**

A solar eclipse occurs when the Moon blocks sunlight from reaching parts of Earth's surface. There are two or three partial eclipses each year, so you'll have a chance to see one every few years. A total eclipse—where the Sun's light is completely blocked out by the Moon and for a few minutes day turns to night —is much rarer. If you want to see one, you may need to travel!

Appearance of an eclipse

In a partial or annular eclipse, the Moon obscures a gradually enlarging, and then gradually diminishing, portion of the Sun. The process lasts anything from a few minutes up to an hour and a half. As our view of the Sun is blocked it becomes darker; and as the obscured area diminishes, light levels return to normal.

A total eclipse is more spectacular. The Sun's disc is completely blocked—typically for 1–3 minutes but sometimes for 7 minutes or more. In this period of totality it becomes as dark as night, stars appear, temperatures drop, and the Sun's corona, a hot, irregularly shaped part of its atmosphere, surrounds a black disc in the sky.

Viewing an eclipse

Never look at the Sun directly with the naked eye or through sunglasses, binoculars, a telescope, or a camera: its intense light can cause eye damage. Use specially designed eclipse-viewing eyeglasses to view an eclipse before and after totality.

How a solar eclipse works

Eclipses are either partial, total, or annular (when the Moon blocks out all but a narrow ring of the Sun's disc). Annular eclipses occur when the Moon is farther from Earth than average, and therefore looks smaller than normal. In a partial eclipse, viewers across a broad swathe of Earth's surface see the Sun partly hidden for a short time.

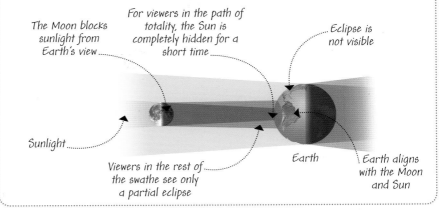

The Moon blocks sunlight from Earth's view

For viewers in the path of totality, the Sun is completely hidden for a short time

Eclipse is not visible

Sunlight

Viewers in the rest of the swathe see only a partial eclipse

Earth

Earth aligns with the Moon and Sun

Calendar dates: total solar eclipses

The table shows when and where you might see a total eclipse, depending on cloud coverage; at the same time a wider area will experience a partial eclipse.

Date	Location
Mar 20, 2015	North Atlantic, Svalbard, Arctic Ocean
Mar 9, 2016	Indonesia, Western and Central Pacific Ocean
Aug 21, 2017	North Eastern Pacific Ocean, United States, Central Atlantic Ocean
Jul 2, 2019	South Pacific Ocean, Chile, Argentina
Dec 14, 2020	Southeast Pacific Ocean, Chile, Argentina
Dec 4, 2021	South Atlantic Ocean, Antarctica
Apr 8, 2024	Central Pacific Ocean, Mexico, United States, Canada
Aug 12, 2026	Greenland, North Atlantic Ocean, Spain

Remember
It is only safe to observe the Sun with the unprotected eye during totality.

2

Build On It

This chapter switches attention to some objects or classes of objects in the night sky that look great through binoculars. These include details on the Moon's surface, such as its seas and craters; a number of sparkling star clusters; some prominent and beautiful nebulae, including the Orion Nebula; the Andromeda Galaxy; bright comets; and the planets Mars and Mercury. You'll learn about a few more constellations—mainly those that contain good objects for binocular viewing. And early in the chapter you'll also find a short guide to choosing and using binoculars.

The Pleiades star cluster in the constellation of Taurus

Choosing and Using **Binoculars**

Binoculars are ideal for taking visual tours across the night sky. If you own a pair for watching wildlife you can probably use them to observe a wide range of astronomical objects, but if you're buying binoculars specifically for astronomy, there are a few things to look out for.

Eyepiece
Eyepiece focusing ring
Prism
Objective lens
Light enters

Standard porro-prism binoculars

Eyepiece with focusing ring
Prisms
Objective lens
Light enters

Compact roof-prism binoculars

Types of binoculars

Binoculars magnify objects and—because they collect more light than the naked eye—make dim objects look brighter. There are two main designs. The porro-prism design has the classic dog-leg shape; they can be wide and bulky but the large aperture of the objective (front) lens means that they tend to have better, brighter optics than the straighter, more compact roof-prism binocular design.

Binoculars are labeled with two numbers, for example 10 x 50. The first number is the magnifying power. High magnification might seem attractive, but can amplify every jolt as you try to hold them steady. A magnification of 10 is a good compromise. The second number is the diameter of the objective lens, and here, the bigger the better. For astronomy, look for an objective lens of at least 50 mm.

Using binoculars comfortably

Its important to be as comfortable as possible. Try reclining on a deck chair, lounger, or waterproof rug on the ground; this will allow you to view the sky at higher angles without straining your neck. Alternatively, sit with your elbows supported by your knees, or rest them on the top of a fence or wall to provide support and stability. Use a tripod for larger, heavier binoculars. Some (pricier) binoculars employ image stabilization technology to reduce the effects of hand shake.

Sitting and placing the elbows on the knees can support the weight of the binoculars and keep them steady

Focusing

Before using your binoculars on the night sky, make sure they are focused on a distant object. The Moon is an obvious choice; if it's not present, try focusing on a cluster of stars. Most binoculars have two focus adjustment wheels—one on the central bar, which moves both eyepieces, and another on one of the eyepieces, which focuses just that eyepiece— the procedure below explains how to focus them. If each eyepiece has its own focusing ring, adjust each eyepiece in turn (closing the other eye), then look through both eyepieces.

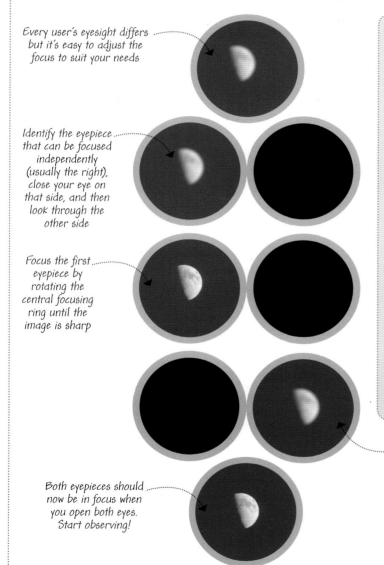

Every user's eyesight differs but it's easy to adjust the focus to suit your needs

Identify the eyepiece that can be focused independently (usually the right), close your eye on that side, and then look through the other side

Focus the first eyepiece by rotating the central focusing ring until the image is sharp

With your other eye open (and the first one closed), adjust the focusing ring on the eyepiece until the image becomes sharp

Both eyepieces should now be in focus when you open both eyes. Start observing!

Messier numbers

In this chapter you will find that some of the sky objects referred to have an alphanumeric descriptor attached. For example, the Pleiades star cluster is also referred to as M45 and a particular galaxy in Centaurus is also referred to as NGC 5128. Descriptors starting with an "M" are called Messier numbers; they refer to the object's placement in a catalog of 110 nebulous-looking sky objects compiled by the French astronomer, Charles Messier. A descriptor starting with "NGC" or "IC" indicates the object's placement in two larger catalogs: the New General Catalog and the Index Catalog.

Observing **the Moon through Binoculars**

Armed with a pair of binoculars, one of the most obvious targets to turn to is the Moon. Through binoculars, vague light and dark areas on the Moon's surface become more obvious and detailed, and you should be able to make out a number of the Moon's largest craters.

Light and shadow

You don't have to wait for the Moon to be full to look at it through binoculars. It can be just as rewarding—in fact, often more interesting—to look at it during one of its crescent, half, or gibbous phases. You can see more detail when the Sun strikes irregularities such as craters at an angle. The dividing line between day and night on the Moon is called the "terminator."

Remember

From the Southern Hemisphere the Moon appears to be upside down—compare the image on p.26. You see the same features, just rotated around 90–180° from their appearance here.

Seas and craters

Using binoculars, try to identify some of the features on the Moon's near side—the face always turned to Earth. You'll see that half of it consists of rough-looking light areas: these are highland regions. The smooth, dark areas were believed by the ancient Greeks to be water, so they named them maria or "seas"; in fact they are lowland basins blasted out by impacts during the Moon's early existence

that later filled with flows of dark lava. Two of the largest are the Mare Serenitatis ("Sea of Serenity") and Mare Imbrium ("Sea of Rains").

Other features to look for include large ray craters, like the Tycho crater, formed when impacting objects hit the surface and sent out splashes of molten rock. It's estimated that the Moon's near side is pitted with about 300,000 craters more than 0.6 miles (1 km) wide.

Mare Crisium

Tycho crater

Mare Imbrium

Plato crater

Mare Serenitatis

Copernicus crater

Mare Crisium

Aristarchus crater

Oceanus Procellarum

Mare Nectaris

Mare Tranquillitatis

Tycho crater

A Quick Guide to **the Moon**

The Moon is much smaller than Earth, with 26 percent of Earth's diameter and only 1.2 percent of its mass. However, when full, it is the brightest object in the night sky, and its gravity influences Earth's oceans. The Moon's surface is a dry, gray, lifeless place, covered in countless impact craters, with a negligible atmosphere.

Orbit

The Moon has a slightly elliptical orbit—at its closest to Earth it is about 10 percent closer than at its farther point. It takes 27.32 days to spin on its axis, the same as the time it takes to orbit Earth. As a result one side of the Moon, its "near side," always faces Earth; no matter where you live, you'll see the same face.

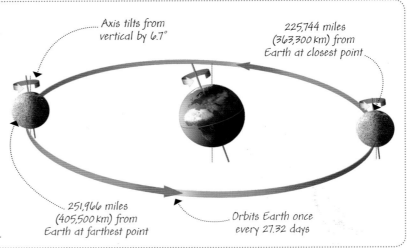

Axis tilts from vertical by 6.7°

225,744 miles (363,300 km) from Earth at closest point

251,966 miles (405,500 km) from Earth at farthest point

Orbits Earth once every 27.32 days

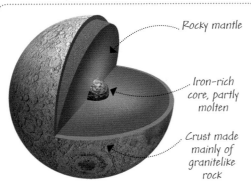

Rocky mantle

Iron-rich core, partly molten

Crust made mainly of granitelike rock

Structure

Like Earth, the Moon has an iron-rich core, a mantle, and a crust, but the core is relatively small. The Moon's atmosphere is extremely thin, with a total mass of less than 24.6 tons (25,000 kg).

Human exploration

Twelve American astronauts (on six separate missions) visited the Moon between 1969 and 1972. They explored limited areas of its surface, took samples of Moon rocks, and left their footprints in the lunar soil.

Moon profile

The Moon is Earth's only natural satellite. With a diameter of 2,159 miles (3,475 km) it is the Solar System's 5th largest planetary satellite.

Average distance from Earth
239,477 miles (385,400 km)

Orbits Earth once every
27 days 7 hours 43 minutes

Rotates once every
27 days 7 hours 43 minutes

Greatest apparent magnitude
-12.9 (see p.51)

Farthest from Earth
251,966 miles (405,500 km)

Nearest to Earth
225,744 miles (363,300 km)

Greatest angular size in sky
34.1 arcminutes (see p.12)

Age
4.5 billion years

Size comparison

Moon

Earth

Heavily cratered surface

Surface temperature varies from up to 257°F (125°C) during the day down to -274°F (-170°C) at night

Observing **Taurus**

An ancient zodiacal constellation, Taurus has been recognized since Babylonian times and is visible from both hemispheres. Its attractions include two large open star clusters—the Hyades and Pleiades—which are excellent targets for binocular viewing.

Taurus, the Bull

In the classic constellation figure, only the front half of a bull is represented. An impressively bright red star, Aldebaran, marks one eye; both the Hyades and more compact Pleiades are easy to find once you have identified it. The Taurid meteors, visible every November, appear to originate just south of the Pleiades.

Remember Taurus is most easily observed during November and December. It is difficult or even impossible to see in May and June because it is too close to the Sun.

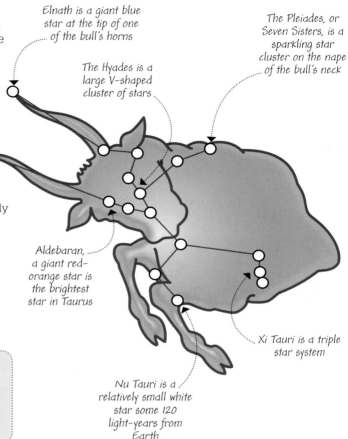

Elnath is a giant blue star at the tip of one of the bull's horns

The Hyades is a large V-shaped cluster of stars

The Pleiades, or Seven Sisters, is a sparkling star cluster on the nape of the bull's neck

Aldebaran, a giant red-orange star is the brightest star in Taurus

Xi Tauri is a triple star system

Nu Tauri is a relatively small white star some 120 light-years from Earth

Northern Hemisphere locators

From midlatitudes (around 40°N), Taurus starts to become visible from early August in the eastern sky—but only in the early morning hours. By December it is prominent and easily seen in the southern sky around midnight.

August 1, 4am
Looking east

December 1, midnight
Looking high in south

April 1, 8pm
Looking west

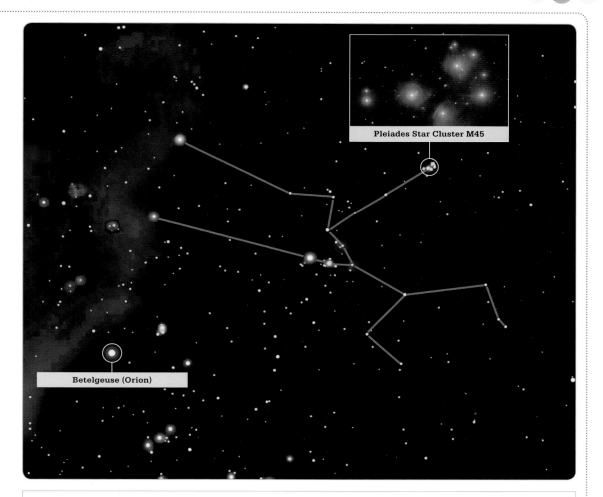

Pleiades Star Cluster M45

Betelgeuse (Orion)

Southern Hemisphere locators

From midlatitudes (around 40°S) Taurus is visible low in the north—and upside down compared to the Northern Hemisphere —in November and December at around midnight. By February it is in the northwest earlier in the evening.

October 1, 2am
Looking northeast

December 1, midnight
Looking north

February 1, 10pm
Looking northwest

A Quick Guide to **Star Clusters**

Large associations of stars that share a common origin are known as clusters. Two types have been identified in the Milky Way Galaxy: "open clusters" are loose groupings of dozens or hundreds of relatively young stars, and "globular clusters" are tight-knit balls of up to a million stars, usually of ancient origin.

Clusters in the Milky Way

Open clusters occur along the Milky Way's spiral arms, often near the nebulae in which they formed. Each star follows its own orbit around the galactic center, so over millions of years the clusters tend to disperse. Globular clusters are permanently bound by the gravity of their close-packed stars. They orbit above and below the plane of the galactic disk.

A few globular clusters orbit close to the central bulge

Globular clusters move in long, elliptical orbits outside the galactic disk

Open clusters are found in the galaxy's spiral arms

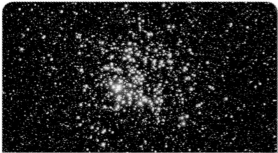

Open star cluster

Irregular in shape, open clusters usually contain between 100 and 10,000 stars, all of which formed around the same time. Bright blue stars indicate a young cluster, while stars in older clusters are yellower.

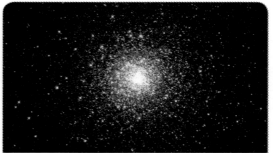

Globular cluster

Large, roughly spherical globular clusters are densely packed with many thousands of ancient red and yellow stars. They are among the oldest stars in the galaxy and help determine the age of the Universe.

Open star clusters M35 (upper left) and NGC 2158 (lower right)

Observing **Cassiopeia**

A distinctive constellation of the northern sky, Cassiopeia is located in a rich area of the Milky Way, full of star clusters that are a fine sight through binoculars. The large W-shape formed by its five main stars—or M-shape depending on your perspective—is one that is very easy to recognize.

Cassiopeia, the Queen

According to Greek mythology, Cassiopeia was an Ethiopian Queen who—because of her vanity—was condemned by Poseidon to circle the north celestial pole forever. In her constellation figure Cassiopeia is classically depicted in a sitting position—and in this representation she is seen admiring herself in a mirror as she endlessly circles the northern sky. Four of the five stars that define Cassiopeia's figure are of roughly equal brightness; the fifth (known as Segin) is slightly fainter.

Remember If you play your binoculars all around Cassiopeia, you'll come across many beautiful star clusters. When doing this you are looking in the direction of the Milky Way, which is where most of the gas and dust that creates star clusters resides.

Segin, a blue-white giant, marks one of the Queen's shins

Tsih, or Gamma Cassiopeia, sometimes erupts to become the brightest star in the constellation

A colorful star cluster, M103, lies just below Cassiopeia's right knee

Shedir, usually the brightest star in Cassiopeia, is an orange giant

---

Northern Hemisphere locators

For an observer at midlatitudes (around 40°N), Cassiopeia is visible in the northern sky for most of the night every night—but the constellation does sometimes partly dip below the horizon, for example at around 2am in March.

June 1, 3am
Looking northeast

October 1, midnight
Looking high in north

February 1, 9pm
Looking northwest

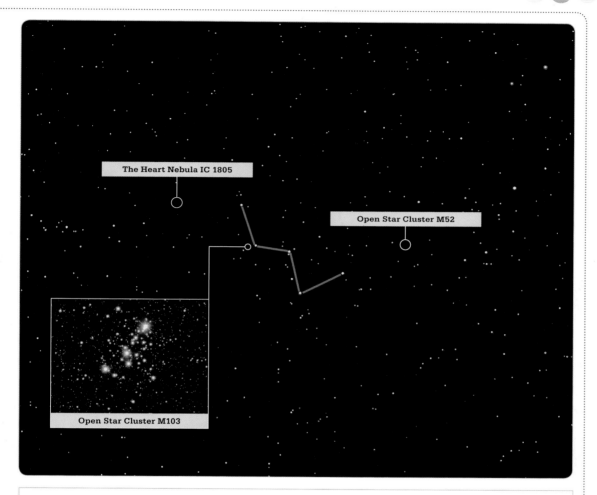

The Heart Nebula IC 1805

Open Star Cluster M52

Open Star Cluster M103

The Heart Nebula IC 1805

This wispy star-forming cloud of dust and gas within the Milky Way Galaxy can be found below Cassiopeia's legs. The nebula is named for its roughly heart-shaped outline, and is about 7,500 light-years away from Earth.

Observing **Hercules**

The large constellation of Hercules represents the hero of Greek myth who undertook 12 labors. Within it you can find a magnificent star cluster known as the Hercules Globular Cluster. This is a fine object to examine through binoculars.

Hercules, the Strong Man

Although it is most easily viewed from the Northern Hemisphere, parts or all of Hercules may also be seen from Southern Hemisphere locations. The figure is arranged around the Keystone—a quadrilateral asterism—and shows Hercules with a club in one hand and the multiple severed heads of Cerberus, a monstrous dog, in the other. Look for the Keystone to locate the constellation.

Remember The Hercules Globular Cluster (M13), situated between two stars in the Keystone, can be seen with the naked eye; through binoculars it looks like a fuzzy ball. It is 25,000 light-years away from Earth and contains over 300,000 closely packed stars.

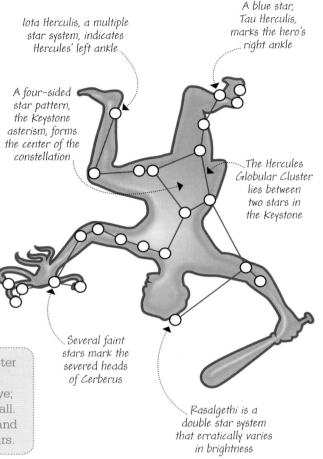

Iota Herculis, a multiple star system, indicates Hercules' left ankle

A blue star, Tau Herculis, marks the hero's right ankle

A four-sided star pattern, the Keystone asterism, forms the center of the constellation

The Hercules Globular Cluster lies between two stars in the Keystone

Several faint stars mark the severed heads of Cerberus

Rasalgethi is a double star system that erratically varies in brightness

Northern Hemisphere locators

From midlatitudes Hercules may be seen during late winter in the east, but only in the early morning hours. By June it is high in the south or overhead at midnight. On fall evenings Hercules is visible in the western sky.

February 1, 3am
Looking east

June 1, midnight
Looking high in south

October 1, 9pm
Looking west

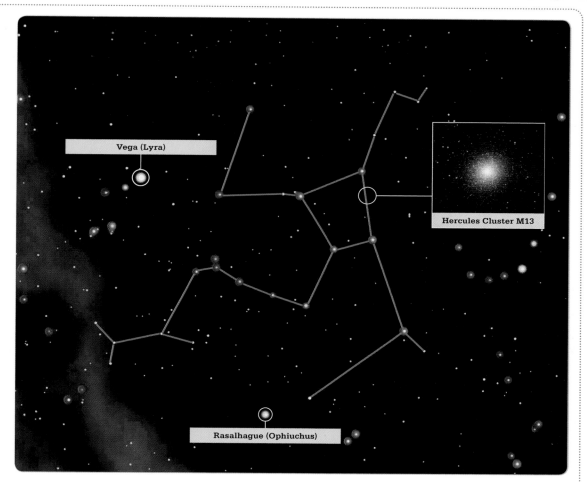

Vega (Lyra)

Hercules Cluster M13

Rasalhague (Ophiuchus)

Southern Hemisphere locators

From midlatitudes, most of Hercules can be seen above the northern horizon around midnight during June. It appears in the same place during July evenings, or in the early morning hours throughout May.

May 1, 4am
Looking north

June 1, midnight
Looking north

July 1, 8pm
Looking north

Observing **Centaurus**

This large constellation contains several objects of interest, including our nearest star neighbors and a massive globular star cluster, Omega Centauri—an excellent target for binoculars. Unfortunately Centaurus is blocked off from the sight of most Northern Hemisphere skywatchers.

Centaurus, the Centaur

The figure depicted is a mythical being with his feet embedded in the Milky Way. Its brightest star, known as Alpha Centauri or Rigil Kentaurus, is a system of three gravitationally bound stars. Other than the Sun, these are the closest stars to Earth—around 4.2 light-years away. The faintest of the three, Proxima Centauri, is individually the closest.

> **Remember** To the naked eye Omega Centauri looks like a smudged star. In fact, it is a globular star cluster, the largest in the Milky Way. More than 12 billion years old, it holds over a million stars—and is definitely worth admiring through binoculars.

Several faint stars mark the centaur's head and arms

Omega Centauri, a giant star cluster, lies close to the center of the constellation

Hadar, or Beta Centauri, is the tenth brightest star in the night sky

The dazzlingly bright Alpha Centauri, a multiple star, marks one hoof of the centaur

Centaurus A galaxy NGC 5128

This unusual galaxy in the centaur's back is just visible through binoculars. The dark band of dust that cuts across is thought to result from a merger of two earlier galaxies. It is 13 million light-years away, with a massive black hole at its center.

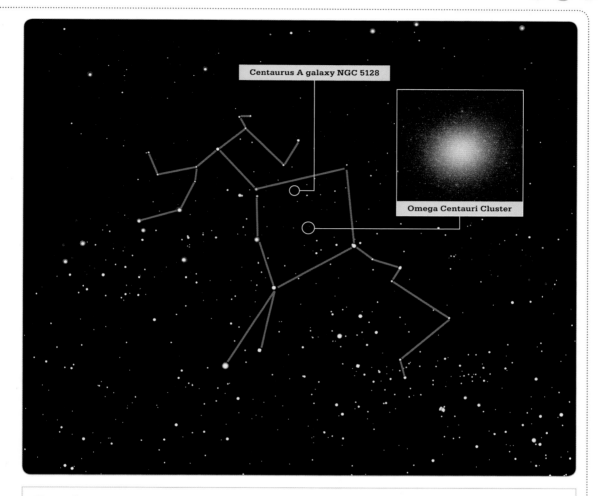

Centaurus A galaxy NGC 5128

Omega Centauri Cluster

Southern Hemisphere locators

From midlatitudes Centaurus is visible for part of the night in most months of the year. Optimal viewing times are March to May, when it is nearly overhead at midnight, or from June to August in the southwestern evening sky.

December 1, 3am
Looking southeast

April 1, midnight
Looking high in south

August 1, 9pm
Looking southwest

Observing **Mars**

Mars is fairly easy to see. It varies considerably in brightness, but it is always visible to the naked eye when present, shining with a reddish light. Through binoculars it appears as a distinct disk.

Tracking Mars

The easiest way to spot Mars is to look for it at around the time of opposition. The chart below shows the months up to 2030 when Mars is at, or close to, opposition. For each month, the constellation that Mars is "in" at the time is also given—so in May 2016 Mars is in Scorpius, between the scorpion's claws.

To locate Mars during opposition, just look into the southern sky (from the Northern Hemisphere) or overhead at about midnight; if you are in the Southern Hemisphere, look north. You should see Mars as one of the brightest objects in view.

Tip Spot Mars at times other than opposition, find out which constellation it is in, then use a planisphere or "app" to work out when and where to see that constellation. This should help you spot Mars too.

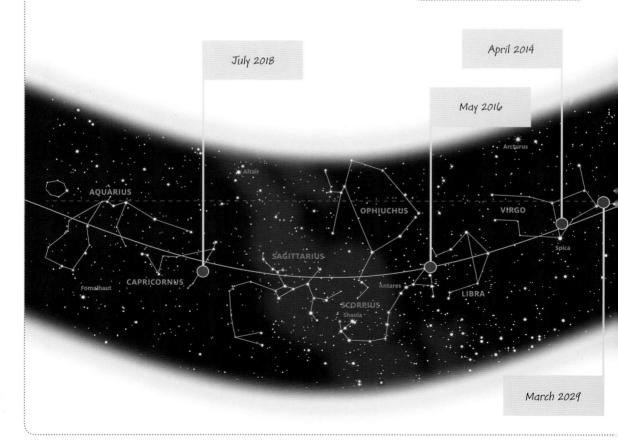

July 2018

April 2014

May 2016

Arcturus

Altair

AQUARIUS

OPHIUCHUS

VIRGO

Spica

SAGITTARIUS

CAPRICORNUS

Fomalhaut

Antares

LIBRA

SCORPIUS

Shaula

March 2029

The view from Earth

Mars takes 687 days (just under 2 years) to orbit the Sun. Earth takes one year, so every 25–26 months it overtakes Mars "on the inside." This is known as "opposition"; for about 6 weeks Mars is at its brightest and closest to Earth—and easiest to see. You can't see Mars when it is on the other side of the Sun from Earth ("conjunction"), which also happens every 25–26 months.

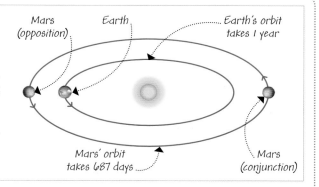

Mars (opposition)

Earth

Earth's orbit takes 1 year

Mars' orbit takes 687 days

Mars (conjunction)

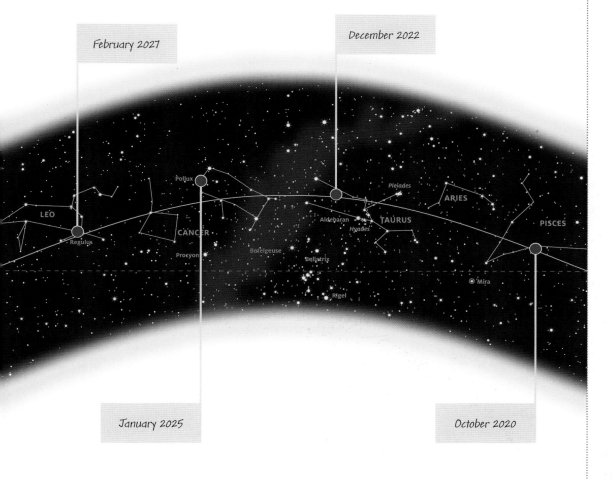

February 2027

December 2022

January 2025

October 2020

LEO

Regulus

CANCER

Procyon

Pollux

Betelgeuse

Bellatrix

Rigel

Aldebaran

Hyades

TAURUS

Pleiades

ARIES

Mira

PISCES

A Quick Guide to **Mars**

Known as the Red Planet due to its rust-red color, Mars is about half the size of Earth and, like our home planet, has seasons and polar ice caps. Evidence from space probes suggests that water once flowed across the surface, but today it is dry and barren, with huge extinct volcanoes, massive canyons, and vast rock-strewn plains.

Orbit

Mars takes 687 days to complete its elliptical orbit of the Sun; a year on Mars is nearly twice as long as a year on Earth. The distance from the Sun varies by up to 26.5 million miles (42.6 million km), and this, combined with its fairly heavily tilted spin axis, causes significant seasonal variation on the planet's surface.

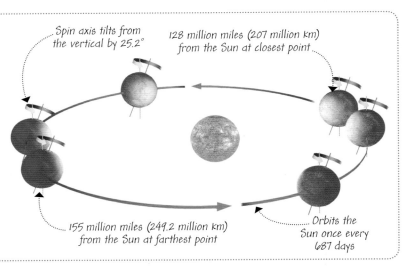

Spin axis tilts from the vertical by 25.2°

128 million miles (207 million km) from the Sun at closest point

155 million miles (249.2 million km) from the Sun at farthest point

Orbits the Sun once every 687 days

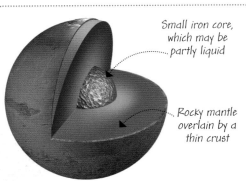

Small iron core, which may be partly liquid

Rocky mantle overlain by a thin crust

Structure

Mars is similar to Earth in structure, with a distinct core, mantle, and crust. Its atmosphere—mainly carbon dioxide—is extremely thin. Winds periodically cause dust storms that cover the whole planet.

Martian gullies

This Martian crater created by an ancient asteroid impact is about 4.4 miles (7 km) across. The narrow gullies eroded into its walls are thought to have been formed by flowing water in the distant past.

Mars profile

Mars is the fourth planet out from the Sun. With a diameter of 4,221 miles (6,792 km), it is the second smallest planet in the Solar System.

Average distance from the Sun
141.6 million miles (227.9 million km)

Orbits the Sun once every
687 days

Rotates once every
24 hours 37 minutes 23 seconds

Greatest apparent magnitude
-2.9 (see p.51)

Farthest from Earth
249.4 million miles (401.3 million km)

Nearest to Earth
33.9 million miles (54.5 million km)

Greatest angular size in sky
25.1 arcseconds (see p.12)

Number of moons
2

Size comparison

Mars

Earth

Polar ice cap of frozen water and carbon dioxide

Rust-red surface with an average temperature of -63°C (-81°F)

Observing the **Orion Nebula**

As you've seen, Orion is one of the most spectacular constellations in the night sky, but nestling within it is a massive glowing cloud of gas and dust called the Orion Nebula—a fantastic sight through binoculars.

Locating the Nebula

Start by finding Orion's Belt with your naked eye (see p.60), and look down from the middle star—or look upward from the middle star if you are observing from the Southern Hemisphere. You should see three stars in a line, forming what is known as Orion's sword; the middle one is rather fuzzy. Aim your binoculars at the sword region and slowly sweep them around until you find the middle "star" of the sword. It will not appear as a single point of light like the other stars, but as a much larger hazy patch. It is not in fact a star at all: this is the Orion Nebula.

Orion from the Northern Hemisphere

Orion from the Southern Hemisphere

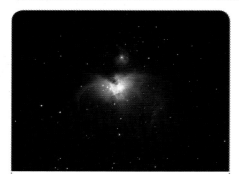

Through a lens

Viewed through binoculars the nebula has a slight greenish or blue-green hue. With a telescope, the green color becomes apparent; the image opposite shows the nebula through a large telescope.

What's in a nebula?

The Orion Nebula is a massive cloud of gas and dust within our own Milky Way galaxy, where new stars and planets are being formed. About 1,350 light-years away from Earth—light from the nebula takes 1,350 years to reach us—and around 24 light-years across, it is the closest region of the Milky Way to Earth in which new stars are being formed. The total mass of gas in the nebula is estimated to be greater than that of 2,000 suns. The nebula glows because of the light from many young stars embedded within it.

Orion Nebula M42 through the Hubble Telescope

A Quick Guide to **Light and Dark Nebulae**

Scattered between the stars of the Milky Way are clouds of gas and dust. Some are dense enough to reflect or block starlight, forming visible structures known as nebulae. The different nebulae are among the most beautiful sights in the galaxy.

Emission nebulae

These brightly colored clouds are illuminated from within by newborn stars. Intense radiation from the stars causes hydrogen atoms in the cloud to ionize (lose electrons); when the electrons recombine with hydrogen nuclei, red light is emitted. Green or blue light indicates that other elements—oxygen, helium, or nitrogen—are present. Emission nebulae include star-forming regions and glowing clouds of debris ejected by dying stars.

Reflection nebulae

Young stars are often surrounded by dust and gas left over from the nebula that gave birth to them. These grains of dust may reflect the bluish light of the hot young stars, forming a blue reflection nebula.

Dark nebulae

Cool clouds of dust and molecular hydrogen can block the light from more distant stars, forming silhouettes known as dark nebulae. These nebulae can collapse under their own gravity and become star-forming regions.

The Bubble Nebula NGC 7635

Observing **Ophiuchus**

This large constellation is rich in globular star clusters, and a fine sight through binoculars. Although it lies partly in the band of sky called the zodiac, it has never been counted as a "sign of the zodiac."

> **Remember** Within the constellation area of Serpens Cauda adjoining Ophiuchus is the Eagle Nebula, or M16, about the size of the full Moon. If you train binoculars on it you may be able to spot a star cluster in M16, which appears as a hazy ball of light.

Ophiuchus, the Serpent-bearer

Ophiuchus is closely associated with the two-part constellation of Serpens— Serpens Caput (Head of the Serpent) and Serpens Cauda (Tail of the Serpent). The figure of Ophiuchus is depicted holding the snake. Its brightest star is Rasalhague, a white giant with a smaller companion, and interesting star clusters are located in different parts—near one foot, around one shoulder (both these areas are in the Milky Way), and also around the center of the constellation.

A beautiful star cluster lies close to the star Celbalrai

Rasalhague

Serpens Caput

Globular clusters M10 and M12 lie in this central area

Delta and Epsilon Ophiuchi are two red giants that appear close in the sky, but are widely separated in space

Serpens Cauda

More globular star clusters lie in front of Ophiuchus' foot

Northern Hemisphere locators

From midlatitudes (around 40°N) most of Ophiuchus can be seen above the northern horizon around midnight during May and June. By September it can be seen low in the southwest earlier in the evening.

March 1, 3am
Looking southeast

June 1, midnight
Looking south

September 1, 9pm
Looking southwest

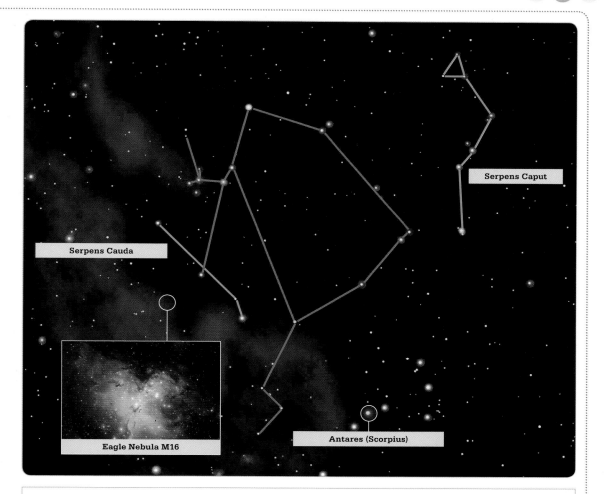

Serpens Caput

Serpens Cauda

Eagle Nebula M16

Antares (Scorpius)

Southern Hemisphere locators

From midlatitudes Ophiuchus is visible from around March in the east, but only in the early hours of the morning. By June it is high in the north or overhead at midnight, while by September it can be seen in the west in the evening.

March 1, 3am
Looking east

June 1, midnight
Looking high in north

September 1, 9pm
Looking west

Observing **Crux**

You met Crux (the Southern Cross) earlier when you used it to "star hop" to neighboring constellations and also to find the south celestial pole. The constellation itself lies in a bright area of the Milky Way and contains some interesting objects for binocular or naked-eye viewing.

Crux, the Southern Cross

Crux is the smallest constellation: from Earth it's just a closed hand's width across. One of its highlights—in addition to the brilliant stars of the cross—is a glittering star cluster, the Jewel Box. Binoculars reveal that it contains several dozen blue-white stars and a contrasting red supergiant. Next to Crux is the Coalsack Nebula, a vast dust cloud.

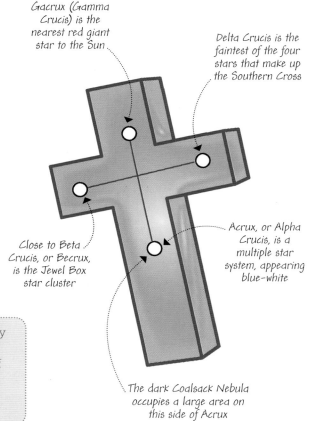

Gacrux (Gamma Crucis) is the nearest red giant star to the Sun

Delta Crucis is the faintest of the four stars that make up the Southern Cross

Close to Beta Crucis, or Becrux, is the Jewel Box star cluster

Acrux, or Alpha Crucis, is a multiple star system, appearing blue-white

The dark Coalsack Nebula occupies a large area on this side of Acrux

Remember Crux is one of the most easily recognized constellations, with a familiar shape—the unmistakable arrangement of its stars is displayed on the national flags of New Zealand, Australia, Samoa, Papua New Guinea, and Brazil.

The Coalsack dark nebula

This large smudgy cloud of interstellar dust is silhouetted against the Milky Way. It spans the width of 12 full Moons—so on dark nights it is easy to spot. Here, the star Acrux is to the right of the nebula, and Beta Crucis is above it.

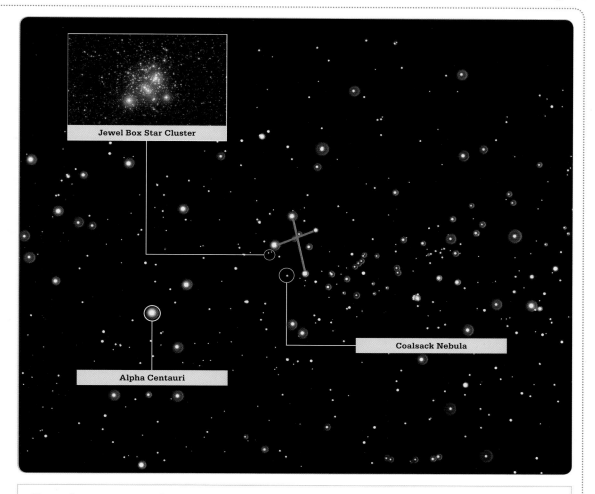

Jewel Box Star Cluster

Coalsack Nebula

Alpha Centauri

Southern Hemisphere locators

From midlatitudes the whole of Crux is visible in the southern sky for much of the night throughout the year. Part of the constellation will sometimes dip below the horizon—for example during November evenings.

December 1, 2am
Looking south

April 1, midnight
Looking high in south

August 1, 10pm
Looking south

Observing **Sagittarius**

This prominent zodiacal constellation lying in the direction of the center of the Milky Way is home to numerous star clusters and bright nebulae, which are a rewarding sight through binoculars. Sagittarius is most easily viewed from the Southern Hemisphere, but on summer nights it can be well worth taking a look at it from the Northern Hemisphere too.

> **Remember** An impressive sight to seek out with binoculars in Sagittarius is the Lagoon Nebula (M8), an emission nebula; it lies above the spout of the Teapot asterism. Farther north lies the Swan—or Omega—Nebula, appearing pale green.

Sagittarius, the Archer

The constellation figure of Sagittarius classically shows a centaur (half human, half horse) with an archer's bow. To find it in the sky, it may help to first identify the Teapot asterism—a pattern of stars at the center of the constellation that resembles a teapot. The area around the top of the teapot is the best place to look with binoculars to see bright nebulae and star clusters.

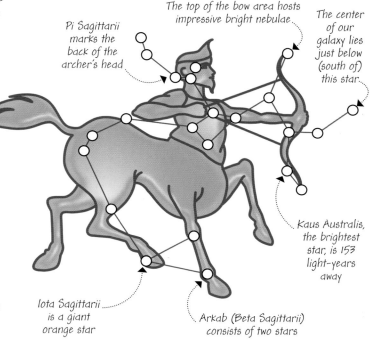

Pi Sagittarii marks the back of the archer's head

The top of the bow area hosts impressive bright nebulae

The center of our galaxy lies just below (south of) this star

Kaus Australis, the brightest star, is 153 light-years away

Iota Sagittarii is a giant orange star

Arkab (Beta Sagittarii) consists of two stars

Northern Hemisphere locators

From midlatitudes (40°N) the constellation of Sagittarius can be viewed between May and September, looking toward the southern horizon. During July, most of the constellation is above the horizon at around midnight.

May 1, 2am
Looking south

July 1, midnight
Looking south

September 1, 10pm
Looking south

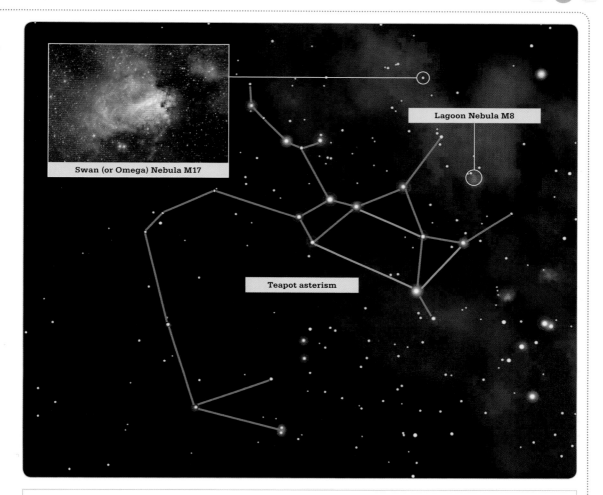

Swan (or Omega) Nebula M17

Lagoon Nebula M8

Teapot asterism

Southern Hemisphere locators

From midlatitudes (40°S), Sagittarius is visible from March in the early hours of the morning, looking east. By July it is high in the north or overhead around midnight. In October it can still be seen in the evening looking west.

April 1, 2am
Looking east

July 1, midnight
Looking high in north

October 1, 10pm
Looking west

Observing **Carina**

A southern constellation that lies partly within the band of the Milky Way, Carina contains a bright nebula, one exceptionally bright star, and several star clusters. Unfortunately, Carina cannot be seen from the Northern Hemisphere unless you are close to the equator.

Carina, the Keel

The constellation of Carina was once part of the larger figure of Argo Navis, the ship of the Argonauts. In the 18th century the keel, sails, and stern were split into Carina, Vela (see p.158), and Puppis. Carina retained many of the original ship's most spectacular features, such as the intense white star Canopus and the bright Eta Carinae Nebula.

> **Remember** Most of Carina's star clusters, which are recommended for binocular viewing, lie near the Carina Nebula, which is easy to spot with the naked eye. An exception is the Diamond Cluster, which lies farther over, toward Canopus.

Puppis, the Stern, is now a separate constellation that borders Carina

Eta Carina, lying within a large, bright nebula, is an unstable star that occasionally flares in brightness

Supergiant Canopus, the second brightest star in the sky, sits 310 light-years from Earth

The single star depicted here is actually a cluster of young stars known as the Southern Pleiades

Eta Carinae Nebula NGC 3372

Carina's centerpiece is a large bright cloud of gas and dust 7,500 light-years away. The Eta Carinae Nebula, or Great Nebula, is an important star-forming region in the Milky Way, and an excellent target for binocular viewing.

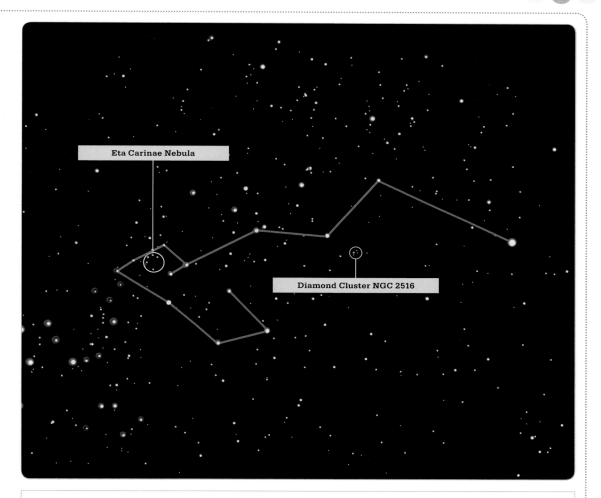

Eta Carinae Nebula

Diamond Cluster NGC 2516

Southern Hemisphere locators

From midlatitudes the constellation of Carina is at least partly visible in the southern sky on every night of the year, although a portion of it does sometimes dip below the horizon—for example, around midnight in August.

October 1, 2am
Looking south

February 1, midnight
Looking high in south

June 1, 10pm
Looking south

Observing **Andromeda**

This northern constellation may not be one that leaps out of the sky, but it does contain some good objects for binocular viewing. Of note is the distant Andromeda galaxy (see p.128) – but first get to know Andromeda itself, and look at some of the closer objects in it.

Andromeda, the Princess

The figure depicts the daughter of the mythical Queen Cassiopeia, who has her own constellation (see p.104). It consists of several strings of relatively faint stars joined to an uneven line of brighter stars running from Andromeda's left foot to her head. Alpheratz (at her head) marks one corner of the Square of Pegasus, a large, easy-to-spot nearby asterism.

Remember The prominent Square of Pegasus surrounds a dark area of sky, so initially you may find it easier to locate than Andromeda itself. Once you've found the Great Square, use the star it shares with the constellation to identify Andromeda.

Almach is a multiple star, with two main components, colored blue and yellow

The Andromeda Galaxy is close in the sky to this star, Nu Andromedae

Alpheratz, the brightest star in Andromeda, forms one corner of the Square of Pegasus

Mirach is a red giant with a diameter 90 times that of the Sun

In Greek mythology, Andromeda is chained to a rock by the sea

Northern Hemisphere locators

At midlatitudes the constellation is visible at some point during the night for most of the year. The best seasons for viewing it are fall and early winter, when Andromeda is high in the south or overhead in late evening.

June 1, 3am
Looking east

October 1, midnight
Looking high in south

February 1, 9pm
Looking west

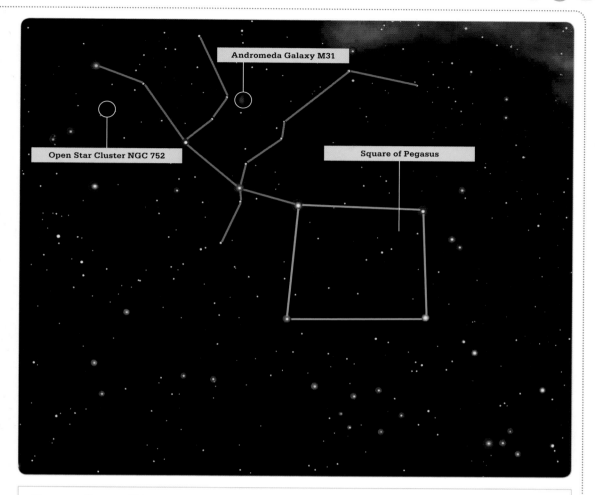

Andromeda Galaxy M31

Open Star Cluster NGC 752

Square of Pegasus

Open Star Cluster NGC 752

This star cluster is found in the area below Andromeda's left leg. It consists of around 70 stars in a loose group, and looks great through binoculars, appearing next to a curving chain of brighter, colored stars to its south.

Observing the **Andromeda Galaxy**

While studying Andromeda and the Square of Pegasus you might like to use your binoculars to seek out a special object—the Andromeda Galaxy—the nearest large galaxy outside our own Milky Way.

Locating the Galaxy

Evenings from October to December, in the Northern Hemisphere or low latitudes in the Southern, are the best times for locating the galaxy. Find Andromeda and the Square of Pegasus just using your eyes. Next, through binoculars, locate the bright star Alpheratz at one corner of the square.

Sweep your binoculars along the figure's body. When you get to the second star, Mirach, turn right along Andromeda's leg toward the "thigh" area. At the second star in the thigh, turn west a little way. You should see a large fuzzy elliptical patch of light—this is the Andromeda Galaxy.

Andromeda from the Northern Hemisphere

Andromeda from the Southern Hemisphere

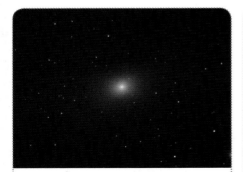

Through binoculars

The stretched-out misty ellipse is larger than the Moon—and this is just the galaxy's central area. The whole galaxy is about twice as big, but its outer parts are too faint to see through binoculars.

Looking through time

On a very dark night with no light pollution the Andromeda Galaxy is the most distant object that can be seen with the naked eye —it's over 2.5 million light-years away, so you see the galaxy as it appeared 2.5 million years ago. A spiral galaxy, like our own Milky Way Galaxy, it contains about a trillion stars and is somewhat larger than the Milky Way, but of comparable mass. Unlike most other galaxies, it is moving toward our own, and in about 3.75 billion years' time, the two are expected to merge.

Andromeda Galaxy M31

A Quick Guide to **Galaxies**

Galaxies are vast, spinning collections of stars, gas, dust, and dark matter, held together by the force of gravity. Astronomers believe there are at least 100 billion galaxies in the Universe. They vary a great deal in size and shape, from dwarfs with just a few million stars to giants with over 100 trillion.

Types of galaxies

Galaxies are classified into just a few main types, depending on the shape we see from Earth. Spiral galaxies have a central hub of stars surrounded by curving spiral arms. In barred spiral galaxies, a straight bar runs across the center, joining the spiral arms. Elliptical galaxies are simple ball shapes, and irregular galaxies have no clear shape.

Elliptical galaxy

Spiral galaxy

Barred spiral galaxy

Irregular galaxy

Spiral galaxy NGC 4414

Active galaxies

Astronomers believe that most galaxies revolve around a central supermassive black hole. Most black holes appear to be dormant, but in certain galaxies—"active" galaxies—matter is falling into the black hole. As it spirals inward, the matter is torn apart by gravity and whirls around the galactic heart to form a ring of superheated dust and gas. Two vast jets of particles blast out from this ring, traveling close to the speed of light and stretching thousands of light-years into space. Active galaxies are classified into various types according to their appearance and the mix of radiation they emit.

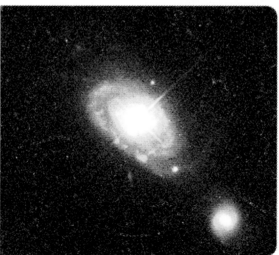

Observing **Leo**

The zodiacal constellation of Leo is one of the easiest to recognize, because its outline really does bear a resemblance to a lion. Highlights for binocular viewing include a few more galaxies as well as some of Leo's brighter stars.

Leo, the Lion

The constellation figure of Leo depicts a crouching lion. At the front of the figure, an asterism called the Sickle traces the lion's chest and head. Good objects for binocular viewing include several stars of the Sickle and a region around the lion's right leg that contains three faint spiral galaxies. The Leonid meteors radiate from the region of the Sickle every November (see p.74).

> **Remember** The bright star Regulus in Leo can be found by "star hopping" from the Big Dipper (see p.46). Once you've found it, you should also be able to see the Sickle asterism, which looks like a backward question mark.

Algieba is a binary star system consisting of two giant orange stars

Binoculars reveal that Zeta Leonis consists of three stars

Just below and behind this star lie the Leo trio: three spiral galaxies— M65, M66, and NGC 3628

Regulus, Leo's brightest star, is a blue-white giant, 80 light-years away

Northern Hemisphere locators

From midlatitudes (around 40°N) the constellation of Leo becomes visible from early November in the eastern sky— but only in the early morning hours. By March it is prominent high in the south at around midnight.

December 1, 2am
Looking east

March 1, midnight
Looking high in south

June 1, 10pm
Looking west

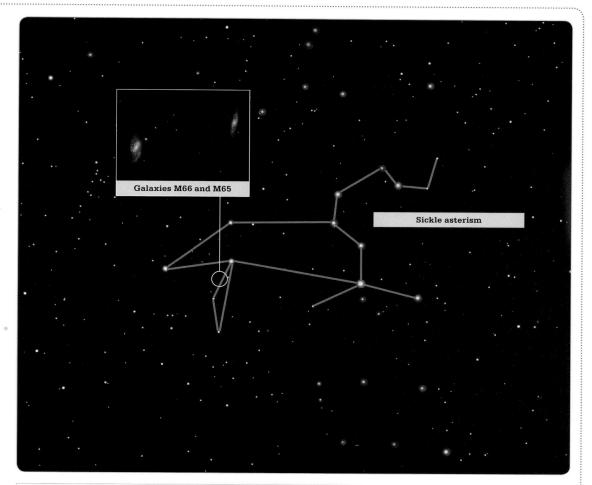

Galaxies M66 and M65

Sickle asterism

Southern Hemisphere locators

From midlatitudes (around 40°S), the constellation can be seen—but upside down—low in the north in February and March at around midnight. By May Leo is low in the northwestern sky earlier in the evening.

January 1, 2am
Looking northeast

March 1, midnight
Looking north

May 1, 10pm
Looking northwest

Observing **Dorado**

From the Southern Hemisphere, it's easy to see a galaxy outside our own: find the constellation of Dorado and look for the Large Magellanic Cloud (LMC), which is an irregular satellite galaxy of the Milky Way. It is visible to the naked eye, but the LMC—as well as a number of other objects in Dorado—is also worth examining through binoculars.

Dorado, the Dolphinfish

The constellation figure depicts a fish—of a type found only in tropical waters—swimming southward. Several of its six main stars are worth looking at through binoculars, but the most interesting object is the nearby Large Magellanic Cloud. Classed as a dwarf galaxy, this is named after the 16th century explorer Ferdinand Magellan. Although some 170,000 light-years away, it spans 20 full Moon diameters in the sky, and at first sight looks like a detached part of the Milky Way. Binoculars bring into view numerous nebulae and star clusters.

Remember To locate Dorado and the Large Magellanic Cloud, first establish the direction of the south celestial pole (see p.44). The "mouth" of the fish and the LMC lie roughly between the pole and Canopus, the brightest star in this part of the sky.

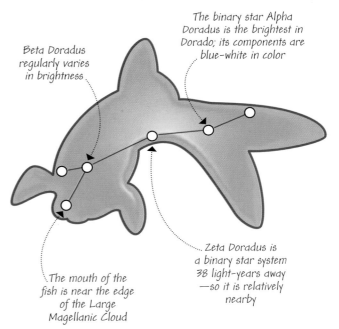

The binary star Alpha Doradus is the brightest in Dorado; its components are blue-white in color

Beta Doradus regularly varies in brightness

The mouth of the fish is near the edge of the Large Magellanic Cloud

Zeta Doradus is a binary star system 38 light-years away —so it is relatively nearby

Tarantula Nebula NGC 2070

This star-forming nebula is near one edge of the Large Magellanic Cloud. At its heart, a cluster of newborn stars can be seen with binoculars. Loops of gas in the nebula resembling a spider's legs inspired its popular name.

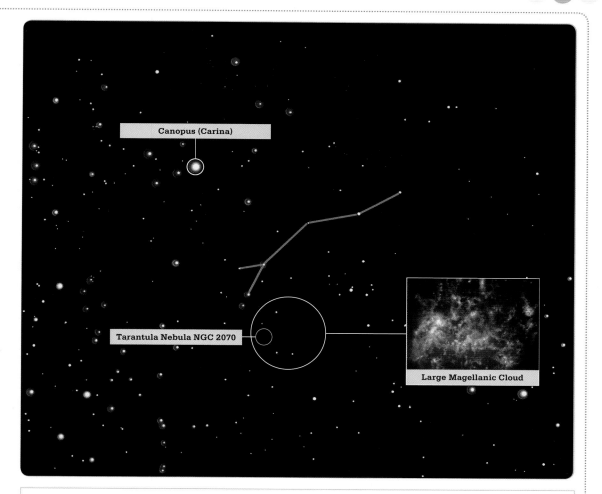

Canopus (Carina)

Tarantula Nebula NGC 2070

Large Magellanic Cloud

Southern Hemisphere locators

From midlatitudes (around 40°S) the constellation of Dorado is at least partly visible in the southern sky every night of the year. Parts of it sometimes dip below the horizon—for example during June and July evenings.

August 1, 2am
Looking southeast

December 1, midnight
Looking high in south

April 1, 10pm
Looking southwest

Observing **Mercury**

Mercury, the closest planet to the Sun, is a tricky object to observe because it never strays far from the Sun in the sky. But from time to time it can be seen as a pinprick of light near the horizon at sunset or dawn. Use of binoculars may reveal its phases.

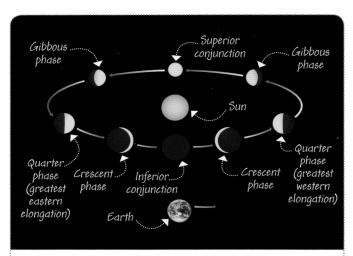

The view from Earth

Mercury orbits the Sun much faster than Earth and roughly every 116 days overtakes Earth "on the inside." At the point of each overtake Mercury moves between Earth and the Sun (inferior conjunction) and cannot be seen. But within a few weeks, and for a few weeks thereafter, it is visible low in the east before dawn (morning apparitions). Subsequently for a few weeks it moves out of view to the other side of the Sun (superior conjunction). The planet then reappears for a few more weeks, this time low in the west just after sunset (evening apparitions). Finally, we lose sight of Mercury as it completes its orbit and overtakes Earth again.

Tip The chances of an observation depend on your latitude, the time of year, and other factors related to Mercury's orbit. But in a clear sky there's a reasonable chance you will see Mercury as a small bright spot very close to the horizon. Try not to confuse it with Venus, which is much brighter and more dazzling.

LOOK EAST before dawn (morning apparitions)	
Date	**Greatest elongation**
Mar 2014	Mar 14
Jul 2014	Jul 12
Oct–Nov 2014	Nov 1
Feb 2015	Feb 24
Jun 2015	Jun 24
Oct 2015	Oct 16
Feb 2016	Feb 6
May–Jun 2016	Jun 5
Sep–Oct 2016	Sep 28
Jan 2017	Jan 19
May 2017	May 17
Sep 2017	Sep 12
Dec 2017–Jan 2018	Jan 1
Apr–May 2018	Apr 29
Aug–Sep 2018	Aug 26
Dec 2018	Dec 15
Apr 2019	Apr 11
Aug 2019	Aug 9
Nov–Dec 2019	Nov 28
Mar 2020	Mar 24
Jul 2020	Jul 22
Nov 2020	Nov 10
Mar 2021	Mar 6
Jun–Jul 2021	Jul 4
Oct 2021	Oct 25
Feb 2022	Feb 16
Jun 2022	Jun 16
Oct 2022	Oct 8
Jan–Feb 2023	Jan 30

LOOK WEST after sunset (evening apparitions)	
Date	**Greatest elongation**
Jan–Feb 2014	Jan 31
May 2014	May 25
Sep 2014	Sep 21
Jan 2015	Jan 14
May 2015	May 7
Aug–Sep 2015	Sep 4
Dec 2015–Jan 2016	Dec 29
Apr 2016	Apr 18
Aug 2016	Aug 16
Dec 2016	Dec 10
Mar–Apr 2017	Apr 1
Jul–Aug 2017	Jul 29
Nov 2017	Nov 23
Mar 2018	Mar 15
Jul 2018	Jul 11
Oct–Nov 2018	Nov 6
Feb–Mar 2019	Feb 27
Jun 2019	Jun 23
Oct 2019	Oct 20
Feb 2020	Feb 10
May–Jun 2020	Jun 4
Sep–Oct 2020	Oct 1
Jan 2021	Jan 24
May 2021	May 17
Sep 2021	Sep 14
Jan 2022	Jan 7
Apr–May 2022	Apr 29
Aug–Sep 2022	Aug 27
Dec 2022	Dec 21
Apr 2023	Apr 11

Mercury (left) with Venus at sunset over Chile, October 2011

Locating Mercury

Mercury is easier to see from tropical and subtropical latitudes, but to view it choose a date between greatest elongation, when it is at its maximum distance from the Sun and most easily spotted, and about a week later for morning apparitions, or a week earlier for evening apparitions. Once you've located Mercury, try examining it through binoculars. You may see that its disk is not fully illuminated but rather half-illuminated or even in a slightly gibbous phase, because Mercury has phases like those of the Moon.

A Quick Guide to **Mercury**

Mercury is one of the more extreme worlds in the Solar System. Due to a unique relationship between its orbit and rate of rotation, most parts of its surface are alternately scorched by the Sun, then plunged into darkness and extreme cold for months at a time. The surface is covered in impact craters and some ancient lava flows.

Orbit

Mercury's 88-day orbit is highly elliptical: at its greatest distance from the Sun it is more than 50 percent farther away than at its closest point. Mercury spins on its axis once every 58.7 Earth days, so it makes three spins for every two orbits. As a result, on Mercury 176 Earth days pass between one sunrise and the next.

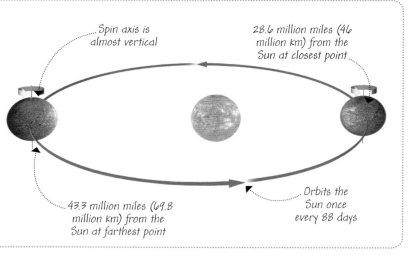

Spin axis is almost vertical

28.6 million miles (46 million km) from the Sun at closest point

43.3 million miles (69.8 million km) from the Sun at farthest point

Orbits the Sun once every 88 days

Structure

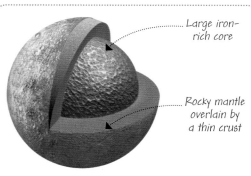

Large iron-rich core

Rocky mantle overlain by a thin crust

Mercury's density is known to be high, so its core must be relatively large and iron-rich. Around the core are a rocky mantle and crust. It is too small and (periodically) hot to retain a significant atmosphere.

Caloris Basin

The circular area with a yellowish hue in this color-enhanced image is known as the Caloris Basin. This wide depression is about 930 miles (1,500 km) across, and was created by an ancient asteroid impact.

Mercury profile

Mercury is the closest planet to the Sun. With a diameter of 3,032 miles (4,879 km) it is also the smallest planet in the Solar System.

Average distance from the Sun
36 million miles (57.9 million km)

Orbits the Sun once every
88 days

Rotates once every
58 days 15 hours 36 minutes

Greatest apparent magnitude
-1.9 (see p.51)

Farthest from Earth
138 million miles (221.9 million km)

Nearest to Earth
48 million miles (77.3 million km)

Greatest angular size in sky
13 arcseconds (see p.12)

Number of moons
0

Size comparison

Mercury

Earth

Surface temperature varies from up to 806°F (430°C) during the day down to -292°F (-180°C) at night

Heavily scarred and cratered surface

Observing **a Comet**

The appearance, for a few weeks or months, of a bright comet in the night sky is a fairly rare astronomical treat. When one of these exotic visitors does appear, spending an hour or two observing it, whether with binoculars or just the naked eye, can be a rewarding experience.

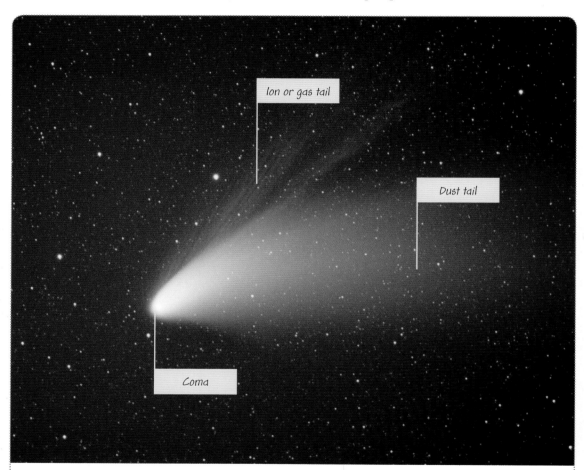

Ion or gas tail

Dust tail

Coma

When to observe a comet

Comets are chunks of ice, dust, and rock that orbit the Sun. As they draw close to the Sun, some develop bright clouds of material around themselves and streaming tails, which makes for a fine sight in the sky. The bad news is that spectacular or "Great" comets are infrequent and unpredictable: there are perhaps one or two per decade. A few return regularly and reliably put on a good show, but the most famous—Halley's Comet—will not be seen again until 2061. Faint comets are much more common, but a "Great" comet will be easier to spot. If one is reported in the media, make sure you see it, even if it means rising at 3am.

Observation technique

A comet does not streak across the sky like a shooting star (see p.74). Look instead for a fuzzy area of light—the coma—which is stationary against the starry background; over several nights you will see that it moves slowly across the starscape. One or more wispy tails may extend from the coma. It is worth observing a bright comet both with the naked eye—to admire the full sweep of the tail(s)—and also through binoculars, which allow you to see extra subtle details in the tails, such as kinks, clumps, and streamers.

Comet Hale-Bopp, July 1997

A Quick Guide to **Comets**

Comets are spinning chunks of ice, dust, and rock. They are anything from about 328 ft (100 m) to 25 miles (40 km) across, and generally occupy remote parts of the Solar System. When a comet enters the inner Solar System, it can become a noticeably bright object in the sky.

Cometary orbits and tails

Comets seen in the inner Solar System are either regular visitors—returning at intervals of anything from a few years to thousands of years—or "once only" tourists. All follow stretched elliptical orbits around the Sun. As a comet nears the Sun, the Sun's heat causes it to release gas, dust, and ice particles into space. These create a bright, fuzzy cloud around the comet's nucleus (central solid part) called the coma. Radiation and streams of particles emitted by the Sun also push on the cloud of ejected material, typically forming two comet "tails." One consists of dust and the other, usually fainter and straighter, consists of ions (charged gas particles).

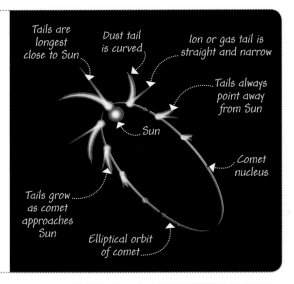

Tails are longest close to Sun

Dust tail is curved

Ion or gas tail is straight and narrow

Tails always point away from Sun

Sun

Comet nucleus

Tails grow as comet approaches Sun

Elliptical orbit of comet

Comet Pan-STARRS

Named for the place it was discovered, this comet put on a little show for skywatchers in the early months of 2013, although it never became very bright. It is not expected to reappear for thousands of years—if ever.

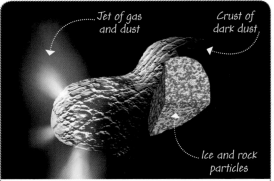

Jet of gas and dust

Crust of dark dust

Ice and rock particles

Nucleus of a comet

A comet's nucleus has an irregular shape and consists of a mixture of ice, dust, and rock particles, with a thin surface layer of dust. It is warmed by the Sun, so jets of gas and dust start to spew from its surface.

Comet McNaught, also called the Great Comet of 2007

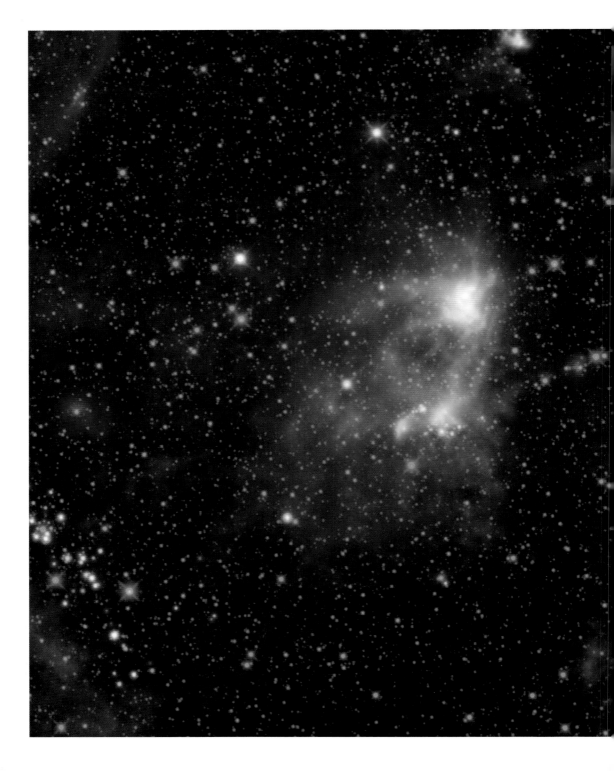

3

Take It Further

The main focus of this final chapter is on some fascinating but rather faint objects in the night sky that you need a telescope to observe clearly. They include Saturn's rings; Jupiter's four largest moons; some planetary nebulae (remains of red giant stars); more galaxies outside our own; and supernova remnants such as the Crab Nebula. Once again, you'll meet a handful of new constellations—this time they contain good objects for telescopic viewing. Also included are guides to choosing and using a small telescope, and the best and safest ways to use equipment if you want to observe the Sun. All these areas are best tackled once you have built up some experience in naked eye and binocular astronomy.

Star-forming nebula (yellow, center) and Tycho's supernova remnant (red bubble at top right)

Choosing and Using **a Small Telescope**

As your interest in astronomy develops, at some point you may want to start using a telescope. This will open up the night skies to new levels of observation, especially for viewing fainter or more indistinct objects such as the moons of Jupiter, Saturn's rings, and some types of nebulae.

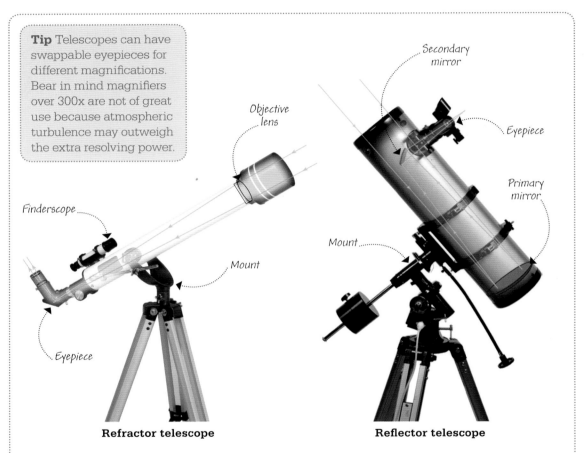

Tip Telescopes can have swappable eyepieces for different magnifications. Bear in mind magnifiers over 300x are not of great use because atmospheric turbulence may outweigh the extra resolving power.

Objective lens

Finderscope

Mount

Eyepiece

Refractor telescope

Secondary mirror

Eyepiece

Primary mirror

Mount

Reflector telescope

Types of telescope

There are two main designs: refractors collect and focus light using a lens, while reflectors use mirrors. Reflectors tend to be cheaper than refractors and more comfortable to use but also more fragile, and they need more maintenance. A third, more complex design—a catadioptric —uses both lenses and mirrors; it's generally more expensive than an equivalent reflector.

Two important factors to consider when choosing your telescope are its light-gathering ability—which depends on aperture—and its magnifying power. Aperture (the diameter of the main lens or mirror) will affect your ability to view faint objects, so the bigger the better. Magnifying power is governed by several components including the eyepiece.

Mounts

The mount not only supports the telescope but also helps track sky objects. There are two main types. An altazimuth pivots in two directions: altitude (up and down) and azimuth (parallel to the horizon); it requires continuous adjustment because sky objects constantly change both altitude and azimuth. Equatorial mounts take account of how objects cross the sky, and once set up (by aligning to one of the celestial poles), objects can be tracked by turning one of the direction controls.

Tip An attractive but more expensive option is a motorized, computerized "Go-to" telescope. Once aligned to the night sky, the telescope can automatically locate and move to a multitude of different celestial objects, without any further assistance from the observer.

Altitude adjustment

Azimuth (compass direction) adjustment

Declination axis

Right ascension axis

Altazimuth mount

Equatorial mount

Finders

All telescopes, even "Go-to" ones, are hard to aim: the typical field of view is 1°, so locating objects can be difficult. A finder on the side of the telescope will help. There are two types. Finderscopes have a field of view of 5–8° and use crosshairs to center the target. Red dot finders project a small dot onto a transparent piece of plastic; the wider sky is visible, so it is intuitive to use. For both types you need to check the alignment from time to time, by looking at a distant fixed object through the main telescope and then through the finder.

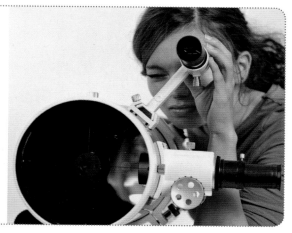

Observing **Saturn**

Saturn is fairly easy to spot in the night sky once you know where to look. It is visible to the naked eye when present, but strong binoculars or a telescope is needed to see its famous ring system.

Tracking Saturn

The easiest way to locate Saturn is to look for it around the time of opposition. The chart below shows all months up to 2030 when Saturn is at, or close to, opposition. For each month, the constellation that Saturn is "in" at the time is also given, so in May 2015 you'll find Saturn in Scorpius.

To spot Saturn around opposition, look into the southern sky (from the Northern Hemisphere) or overhead at about midnight in the month in question; if you are in the Southern Hemisphere, look north. Saturn is one of the brighter objects. Astronomical software or a smartphone "app" may also help you locate Saturn.

Tip Saturn only moves constellations once every 2–3 years. When it's not in opposition, look for the nearest constellation figure. If you work out when and where to see that, you may be able to spot Saturn too.

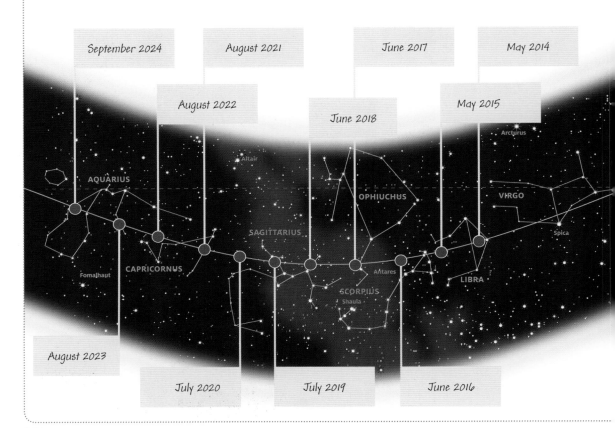

The view from Earth

Saturn takes nearly 30 years to orbit the Sun once. Earth takes just one year, so it overtakes Saturn "on the inside" roughly every 378 days (just over a year). At this time, "opposition," Saturn is at its closest to Earth and at its brightest. Saturn is not visible when it is on the other side of the Sun from Earth ("conjunction"), which also occurs approximately every 378 days.

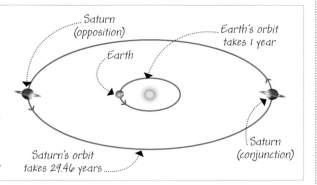

Saturn (opposition)

Earth's orbit takes 1 year

Earth

Saturn's orbit takes 29.46 years

Saturn (conjunction)

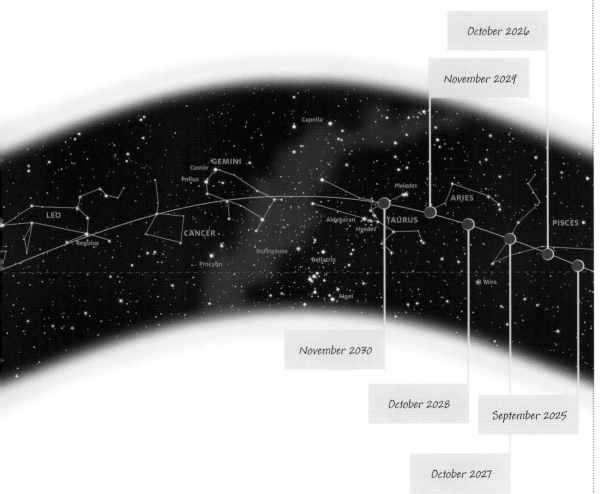

October 2026

November 2029

November 2030

October 2028

September 2025

October 2027

A Quick Guide to **Saturn**

Saturn is a colossal world consisting almost entirely of the lightest chemical elements, hydrogen and helium. Its magnificent rings are composed of lumps of ice that vary in size from dust-sized specks to icy boulders the size of a car; vast in breadth, on average they are just 66ft (20m) thick. Saturn also has a large family of moons.

Orbit

Saturn has a markedly elliptical orbit: its distance from the Sun varies by up to 100.7 million miles (161.9 million km). Saturn tilts heavily as it spins, causing slow seasonal changes as it moves around the Sun. The axial tilt also make the view of Saturn's rings, as seen from Earth, continuously change.

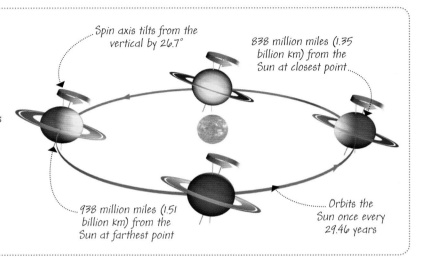

Spin axis tilts from the vertical by 26.7°

838 million miles (1.35 billion km) from the Sun at closest point

938 million miles (1.51 billion km) from the Sun at farthest point

Orbits the Sun once every 29.46 years

Liquid metallic hydrogen and helium

Rock and ice core

Liquid hydrogen and helium

Atmosphere of hydrogen and helium gas

Structure

Saturn's center is thought to be a dense core of rock and ice. Surrounding this are liquid layers, mainly of hydrogen and helium, which gradually merge into an atmosphere of hydrogen and helium gas.

Saturn's moon Titan

The largest of Saturn's many moons, Titan is the only moon in the Solar System with a dense atmosphere; as on Earth, this consists mostly of nitrogen. On Titan's surface are lakes of liquid methane and ethane.

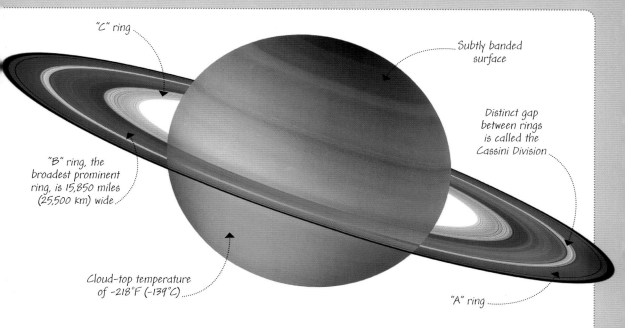

"C" ring

Subtly banded surface

Distinct gap between rings is called the Cassini Division

"B" ring, the broadest prominent ring, is 15,850 miles (25,500 km) wide

Cloud-top temperature of –218°F (–139°C)

"A" ring

Saturn profile

Saturn is the sixth planet out from the Sun. With a diameter of 74,898 miles (120,536 km), it is the second largest planet in the Solar System.

Average distance from the Sun
890.8 million miles (1.43 billion km)

Greatest angular size in sky
20.1 arcseconds (see p.12)

Orbits the Sun once every
29.46 years

Number of moons
62 confirmed as of July 2013

Rotates once every
10 hours 32 minutes

Size comparison

Greatest apparent magnitude
-0.49 (see p.51)

Farthest from Earth
1,030.5 million miles (1,658.5 million km)

Nearest to Earth
742.8 million miles (1,195.5 million km)

Earth

Saturn

Observing **Jupiter's Galilean Moons**

Now that you're using a telescope, try taking a closer look at Jupiter—see p.68 for a reminder of how to locate it in the night sky. You may be able to pick out some banding and possibly the Great Red Spot (see p.70) on the surface, as well as some of its moons. The four largest were spotted in 1610 by the Italian astronomer Galileo Galilei, who was observing Jupiter through an early telescope—use a small modern telescope to repeat his exercise.

The Galilean moons

Jupiter has many moons—67 had been discovered by July 2013—but the four Galilean moons are much bigger than the rest. Galileo's discovery of Io, Europa, Ganymede, and Callisto was of huge scientific significance because it assisted in debunking the standard belief of the time that everything in space revolved around Earth.

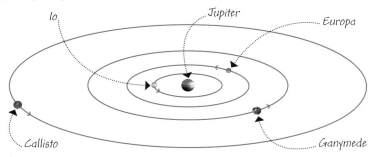

The changing configuration of Jupiter's moons

If you observe the moons over several nights, you'll see that their configuration changes every night. Galileo also noted this and figured out that it must be because the moons orbit Jupiter, doing so at different orbital periods. Io, the closest to Jupiter, orbits in about 1.8 days; Europa in 3.6 days; Ganymede in 7.2 days; while Callisto takes more than 16 days.

Moon movements
The diagram shows how their varied orbits appear to affect the order of the Galilean moons on successive nights. The moons in the top section are (from left to right) Io, Ganymede, Europa, and Callisto.

Locating the Galilean moons

Aim your telescope at Jupiter and look around it for up to four small white starlike dots, of roughly the same size and brightness in a fairly straight line with Jupiter itself. These are the Galilean moons. They may all be on one side of Jupiter but usually you'll see some to either side, and sometimes one or two will be hidden by the planet.

Jupiter and the four Galilean moons (left to right): Callisto, Europa, Io, and Ganymede

Observing **Aquarius**

This large zodiacal constellation can be seen from both hemispheres. It contains a handful of objects that are good targets for a telescope, including two planetary nebulae— glowing shells of gas ejected by red giant stars as they begin to die.

Remember In a clear dark sky, the Helix Nebula is just visible with a small telescope as a pale fuzzy disk, half the width of the full Moon. Around 700 light-years away, it is thought to be the closest planetary nebula to Earth.

Aquarius, the Water Carrier

The figure depicts a young man pouring water from a jar. One planetary nebula, the Saturn Nebula, is located just below a star near the left hand. The other, the Helix Nebula, lies south of the figure. Another notable telescopic object is globular star cluster M2, situated in the vicinity of the head.

A Y-shaped pattern of four stars marks the water jar.

Sadalsuud, a Sun-like star, is the brightest in Aquarius

Just south and east of this star lies the Saturn Nebula, appearing as a greenish disk

Delta Aquarii lies close to the radiant (focal point) of a meteor shower that peaks at the end of July

Northern Hemisphere locators

From midlatitudes (around 40°N) the constellation of Aquarius is visible above the southern horizon around midnight in August and September. By December it is low in the southwest earlier in the evening.

June 1, 3am
Looking southeast

September 1, midnight
Looking south

December 1, 9pm
Looking southwest

Globular Cluster M2

Helix Nebula NGC 7293

Fomalhaut (Piscis Austrinus)

Southern Hemisphere locators

From midlatitudes (around 40°S) Aquarius can be seen from June in the northeast in the early morning. By September it is visible high in the north at midnight, and during December in the northwest in the evening.

June 1, 3am
Looking northeast

September 1, midnight
Looking north

December 1, 9pm
Looking northwest

A Quick Guide to **Planetary Nebulae**

Planetary nebulae have little to do with planets—they are small, typically rather spherical clouds of gas in deep space that can look a bit like planets when viewed through a small telescope. Often rather ghostly in appearance, they are formed when Sun-like stars reach a stage in their life cycles where they shed their outer layers.

Formation of a Planetary Nebula

As a star of similar mass to the Sun uses up the fuel it needs to create energy, it swells into a red giant. The red giant eventually ejects most of its material as a halo of gas— a planetary nebula. The rest of it collapses to form a hot dense star called a white dwarf.

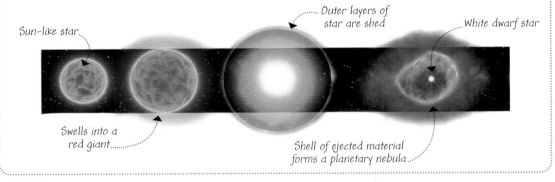

Sun-like star

Swells into a red giant

Outer layers of star are shed

White dwarf star

Shell of ejected material forms a planetary nebula

Cat's Eye Nebula

This nebula, in the constellation of Draco, is relatively bright and visible with a small telescope. Large instruments reveal its complex structure of arc- and bubblelike features and the central white dwarf star.

Owl Nebula

This nebula, also called M97, is a faint object in the constellation of Ursa Major, close to the star Merak (see p.166). A large telescope discerns the dark patches in the nebula that look like an owl's eyes.

Kronberger 61 Nebula (in Cygnus)

Observing **Vela**

The southern constellation of Vela sits partly in the Milky Way and contains several excellent objects for telescopic viewing. They include the Eight-Burst Nebula (a planetary nebula), a globular star cluster, and an open star cluster, as well as Regor, a bright binary star with additional fainter companion stars.

Vela, the Sails

The constellation figure depicts sails that were originally part of the larger constellation of Argo Navis (see Carina, p.124). The objects in Vela recommended for telescopic viewing all lie close to the sails' outline. The Eight-Burst Nebula appears as a pale oval disk with a tiny white dwarf star at its center.

Remember Two of Vela's stars—Kappa Velorum and Koo She (Delta Velorum)—are part of the False Cross asterism, which you may have explored in a "star hopping" exercise (see p.49). This is an excellent way to find Vela in the southern night sky.

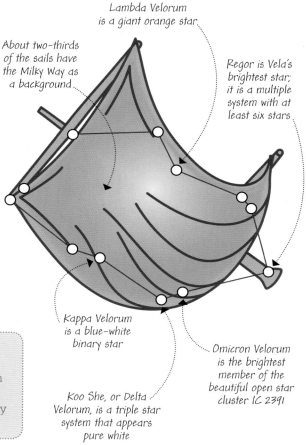

Lambda Velorum is a giant orange star.

About two-thirds of the sails have the Milky Way as a background

Regor is Vela's brightest star; it is a multiple system with at least six stars

Kappa Velorum is a blue-white binary star

Omicron Velorum is the brightest member of the beautiful open star cluster IC 2391

Koo She, or Delta Velorum, is a triple star system that appears pure white

Globular Star Cluster NGC 3201

This star cluster is about 10 billion years old, lies 16,300 light-years away, and contains hundreds of thousands of stars. Some can be seen individually through a telescope. Many are red giants, giving it a slightly orange hue.

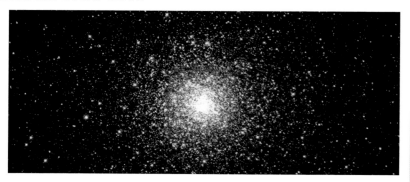

Eight-Burst Nebula

Globular Cluster NGC 3201

Open Star Cluster NGC 2547

Southern Hemisphere locators

From midlatitudes (around 40°S) Vela is visible for part of the night in every month. View from January to March, when it is high in the south or nearly overhead at midnight, or from April to July in the southwestern evening sky.

October 1, 2am
Looking southeast

February 1, midnight
Lookng high in south

June 1, 10pm
Looking southwest

Safely Observing **the Sun**

In addition to all the objects that can be observed at night, the centerpiece of our Solar System—the Sun—can be fascinating to study, but you must be stringent about safety when observing it. You should never look at the Sun with the naked eye, or through any instrument without using special filters.

Projection techniques

The safest method to observe the Sun is to project its image onto a piece of white card. You can do this by directing the Sun's rays through a telescope eyepiece or one side of a pair of binoculars. The projection method is suitable for observing phenomena such as sunspots and solar eclipses; you do not look directly at the Sun so there is no chance of any light or other radiation impinging onto your eyes, which could cause severe damage, or even blindness. You may need to experiment a bit to get a clearly focused image on the card. Start by placing the card about 18 in (50 cm) from the telescope or binocular eyepiece. The farther the card is from the eyepiece, the larger—but fainter—the image will be.

> **Remember** When projecting the Sun, be careful not to put your hand or anything flammable near the telescope or binocular eyepiece: the concentrated sunlight can cause a nasty burn or set something on fire. Never leave a solar telescope—or a telescope with add-on filter—unattended, especially if children are present.

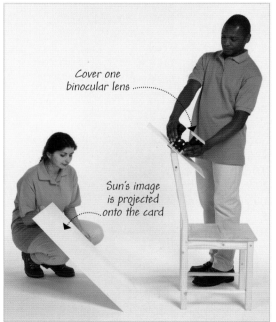

Cover one binocular lens

Sun's image is projected onto the card

Image of Sun focused on card

Telescope projection

Cap the main lens and finderscope. To aim at the Sun move the telescope around until you find the position where it casts the shortest shadow. Uncap the main lens and adjust the eyepiece to sharpen the image on the card.

Binocular projection

Cover one binocular lens so that sunlight can only pass through the other side. Use the binocular lens to direct the Sun's image onto the card. Adjust the viewing card until the Sun's image is in focus.

Add-on filter on front of telescope

Cap finderscope or remove it entirely

Solar telescope with built-in filter

Solar filters and telescopes

An alternative to the projection method is to fit an ordinary telescope with a solar filter. Always adhere to the manufacturers' instructions, and only use add-on filters with the telescopes for which they are designed. Before use, hold the filter up and check that no pinpoints of light penetrate through scratches in its surface. If there are any flaws, reject it. Make sure the filter fits snugly over the telescope and cannot be dislodged. Cap or remove the finderscope, then target the Sun using the "shortest shadow" method. Specially designed solar telescopes incorporating filters and other devices for viewing the Sun are also available.

Solar phenomena to observe

There are a number of phenomena that can be observed on or around the Sun using projection methods or filters:

Solar eclipses—see pp.90–91.

Sunspots—these are darker, cooler regions on the Sun's surface, caused by localized disturbances in the Sun's magnetic field.

Faculae—brighter, more active regions on the Sun's surface.

Solar prominences—huge loops of material from the Sun extending into its atmosphere.

Coronal mass ejections—huge clumps of matter from the Sun that have broken off and are heading into space.

Transit of Mercury—when Mercury passes across the face of the Sun. The next transits occur on May 9, 2016 and November 11, 2019.

A Quick Guide to **the Sun**

The Sun, our local star, is an incandescent ball of extremely hot plasma (ionized gas) with 750 times the mass of all the Solar System's planets combined. At its center, vast amounts of energy are produced—this eventually escapes from the Sun's surface as heat, light, and other forms such as ultraviolet rays.

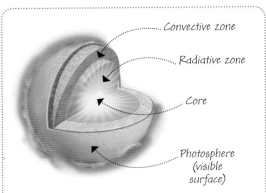

Convective zone

Radiative zone

Core

Photosphere (visible surface)

Structure

The Sun has an extremely hot core where energy is produced by nuclear fusion. It is surrounded by a zone where energy is radiated outward, then a convective zone where hot gas bubbles rise to the surface.

Surface features

Various disturbances occur at the Sun's surface including violent explosions known as solar flares, jets of hot gas or huge loops of plasma shooting up or drifting into the Sun's atmosphere, and solar quakes.

Sunspot cycle

Sunspots are dark, relatively cool regions that appear where hot gas cannot reach the Sun's surface. They emerge in bands on either side of the Sun's equator. The number and pattern of spots varies over approximately 11-year cycles. In each cycle, the bands where spots appear gradually migrate toward the equator. At what are called solar maxima, when there are many spots near the equator, other types of surface disturbances occur. Some of these release vast amounts of energy and electrically charged matter into space.

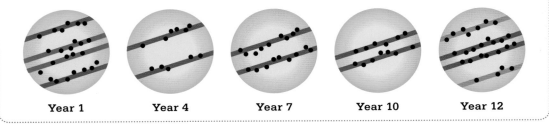

Year 1   Year 4   Year 7   Year 10   Year 12

Sun profile

With a diameter of 869,900 miles (1.4 million km) and a mass 330,000 times that of Earth, the Sun is a stable, medium-sized yellow star.

Distance from center of Milky Way
About 27,000 light-years

Orbits the galactic center every
225–250 million years

Rotates once every
25 days (equatorial region)
34 days (polar region)

Greatest apparent magnitude
-26.7 (see p.51)

Nearest to Earth
91,445,000 miles (147,166,000 km)

Greater angular size in sky
32.7 arcminutes (see p.12)

Age
6 billion years

Number of planets 8

Size comparison

Earth

Sun

Surface mottling is called granulation

Bright, active region, called a facula

Temperature of visible surface is about 9,570°F (5,300°C); it is believed to be around 28.3 million °F (15.7 million °C) at its core

Observing **Virgo**

What the large zodiacal constellation of Virgo lacks in bright star clusters and nebulae, it makes up for in the number of the galaxies it contains—and some of them can be seen with a small telescope. Virgo straddles the celestial equator and can be seen equally well from both hemispheres.

> **Remember** Virgo is most easily observed in the months of March, April, and May. It is difficult or impossible to view the constellation during September and October because it is too close to the Sun.

Virgo, the Virgin

The constellation figure depicts a maiden goddess, identified by some as Demeter. In a limited area of the constellation—50 million light-years away but still worth targeting with a telescope—lies the Virgo cluster of galaxies. Highlights in other regions of this constellation include the Sombrero spiral galaxy (named for its resemblance to a Mexican hat) and the binary star Porrima.

The Virgo cluster of spiral and elliptical galaxies lies on this side of the virgin's body

Vindemiatrix is a giant yellow star

A large Y-shaped pattern of stars identifies this part of the constellation figure

Spica, a blue-white binary star, is the brightest in Virgo

Porrima consists of two near-identical white stars that orbit each other every 169 years

---

Northern Hemisphere locators

From midlatitudes (around 40°N) the constellation becomes visible from about early January in the southeastern sky—but only in the early morning. By April Virgo is prominent in the south at around midnight.

January 1, 3am
Looking southeast

April 1, midnight
Looking south

July 1, 9pm
Looking southwest

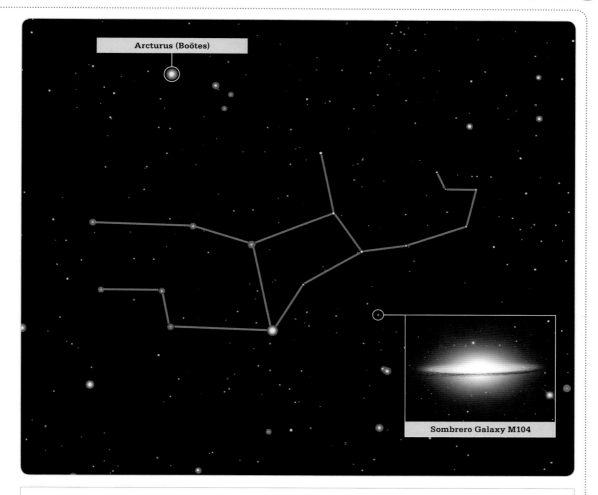

Arcturus (Boötes)

Sombrero Galaxy M104

Southern Hemisphere locators

From midlatitudes (around 40°S) Virgo is visible in the northeast early in the morning in January, or in the northern sky at around midnight in March and April. By July the constellation is in the northwest earlier in the evening.

January 1, 3am
Looking northeast

April 1, midnight
Looking north

July 1, 9pm
Looking northwest

Observing **Ursa Major**

A large northern constellation, Ursa Major is best known as the home of the Big Dipper asterism, which provides a useful signpost to several stars and constellations (see pp.46–47). A lesser known fact is that within the constellation's borders lie some nearby galaxies that may be viewed through an amateur telescope.

Ursa Major, the Great Bear

The constellation figure is an adult bear striding across the northern sky. Try training your telescope on the Pinwheel Galaxy, located near the handle of the Big Dipper, and Bode's Galaxy, nearer the north celestial pole. Also worth viewing are two stars very close to each other in the Big Dipper's handle, Mizar and Alcor.

Remember Bode's Galaxy is one of Ursa Major's best objects to see with a telescope. This large spiral galaxy lies in the region of the bear's head. About one diameter of the Moon away from it is another galaxy that lies edge-on—known as the Cigar Galaxy.

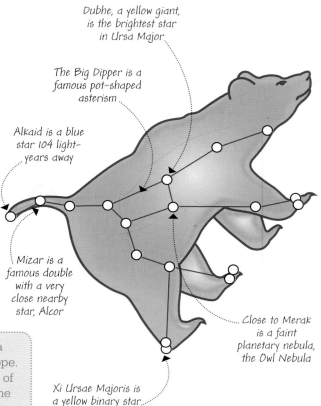

Dubhe, a yellow giant, is the brightest star in Ursa Major

The Big Dipper is a famous pot-shaped asterism

Alkaid is a blue star 104 light-years away

Mizar is a famous double with a very close nearby star, Alcor

Close to Merak is a faint planetary nebula, the Owl Nebula

Xi Ursae Majoris is a yellow binary star

Northern Hemisphere locators

For observers at midlatitudes (around 40°N), Ursa Major is wholly visible at some point during the night all year round. Summer is a good season to view it, when it is prominent in the northwest in the late evening.

December 1, 3am
Looking northeast

April 1, midnight
Looking high in north

August 1, 9pm
Looking northwest

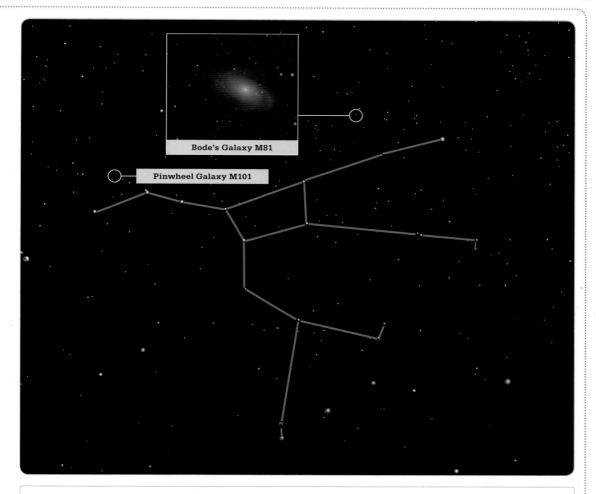

Bode's Galaxy M81

Pinwheel Galaxy M101

Pinwheel Galaxy M101

Around 21 million light-years away, the Pinwheel Galaxy (M101) forms the tip of a triangle with Mizar and Alkaid. It is face-on to Earth and just visible with a small telescope; a larger instrument will reveal the spiral arms.

Observing **Hydra**

The large constellation of Hydra snakes more than a quarter of the way across the sky. Several different object types can be seen through a telescope in this constellation, including a planetary nebula, a spiral galaxy, and an open star cluster.

> **Remember** One of the finest objects to view is star cluster M48, a loose collection of at least 80 stars occupying an area larger than that of the full Moon. Through a telescope these stars can be seen to be blue in color, with some yellow ones mixed in.

Hydra, the Water Snake

Hydra represents the monster that was killed by Hercules in the second of his labors. In the sky, it consists of a small group of stars representing the snake's head, and a dozen further stars that weave to the south of the constellations of Leo and Virgo. The main features of interest for telescopic viewing lie beneath the snake's head, tail, and belly.

A distinctive five-star asterism forms the head of the snake ......

Alphard, an orange giant, is the brightest in Hydra ......

An open star cluster, M48, lies in the area below the snake's head

R Hydrae is a red giant star that varies enormously in brightness ......

Just to the south of Mu Hydrae lies a planetary nebula known as the Ghost of Jupiter

The Southern Pinwheel Galaxy (M83) lies just below the tail of the water snake

Northern Hemisphere locators

Hydra is so big that viewing the whole constellation at one time is difficult. For observers at midlatitudes (around 40°N) however, it can be seen around midnight in early spring, looking from south to west.

April 1, midnight

Looking south   Looking southwest   Looking west

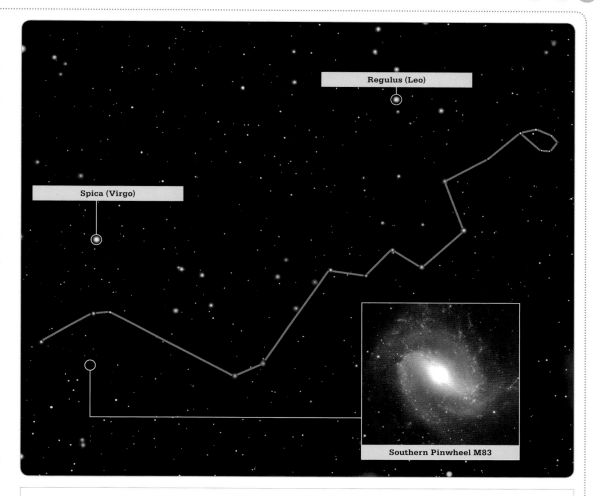

Regulus (Leo)

Spica (Virgo)

Southern Pinwheel M83

Southern Hemisphere locators

From midlatitudes (around 40°S) it should be possible to view the whole constellation of Hydra at around 10pm in mid-February, looking in an area of the sky extending from the north to the southeastern horizon.

February 15, 10pm

Looking north   Looking northeast   Looking east   Looking southeast

Observing a **Supernova Remnant**

Through your telescope you may be able to observe the remains of a gigantic explosion in space, first seen on Earth nearly 1,000 years ago. The Crab Nebula in Taurus is an example of a "supernova remnant."

Locating the Crab Nebula

The best time to observe the Crab Nebula from either hemisphere is on evenings from December to March. First locate Taurus, just using your eyes (see pp.100–101 to remind yourself how to find it). Next, still with the naked eye, locate the two bright stars at the end of the bull's "horns." The one you are interested in is the star at the tip of the bull's southern horn, called Zeta Tauri. Aim your telescope at this star. In the region around it, you should spot a small oval smudge in the sky, with a shape vaguely resembling a crab's shell: this is the Crab Nebula.

Taurus from the Northern Hemisphere

Taurus from the Southern Hemisphere

Remnant in Vela

Other supernova remnants are dotted around the sky. This one, viewable in the southern sky only with a powerful telescope, is the remains of a star that exploded about 12,000 years ago.

A cloud of debris

The Crab Nebula is the remains of a star that exploded some 7,500 years ago. The star was about 6,500 light-years away, so the blast was not seen on Earth until 1054, when Chinese astronomers noted a new bright "star." The remnants are still expanding at the rate of 930 miles (1,500 km) per second, and the cloud is now about 11 light-years across. At its center is a strange object called a pulsar—an exceedingly dense star (a neutron star) that spins rapidly and emits powerful pulses of radiation, including X-rays and radio waves.

Crab Nebula M1 through the Hubble Telescope

A Quick Guide to **Supernovas**

A supernova is the exceedingly violent explosion of a high-mass star at the end of its life. It produces a vast outpouring of radiation, and a bright new light remains in the sky for weeks or months. Supernovas in our own galaxy are rare—the last conspicuous one was observed in 1604.

How supernovas develop

When high-mass stars that are at least 8 times more massive than the Sun run out of fuel, they swell into supergiant stars. Eventually these explode as supernovas and—depending on their original mass—end up as either ultra-dense neutron stars or as black holes.

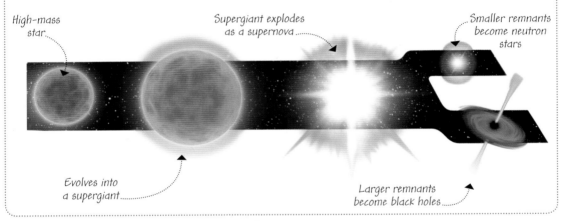

High-mass star

Supergiant explodes as a supernova

Smaller remnants become neutron stars

Evolves into a supergiant

Larger remnants become black holes

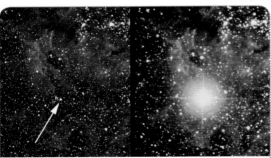

Appearance of a supernova
The most recent supernova visible to the naked eye appeared in 1987 in a nebula outside our galaxy. The images above show how the nebula looked before (left) and after (right) the supernova came into view.

Tycho's supernova remnant
The small red cloud at the top left of this image is the remnant of a supernova first seen on Earth in 1572. Danish astronomer Tycho Brahe noted that the supernova was so bright it was visible in broad daylight.

Supernova remnant Cassiopeia A

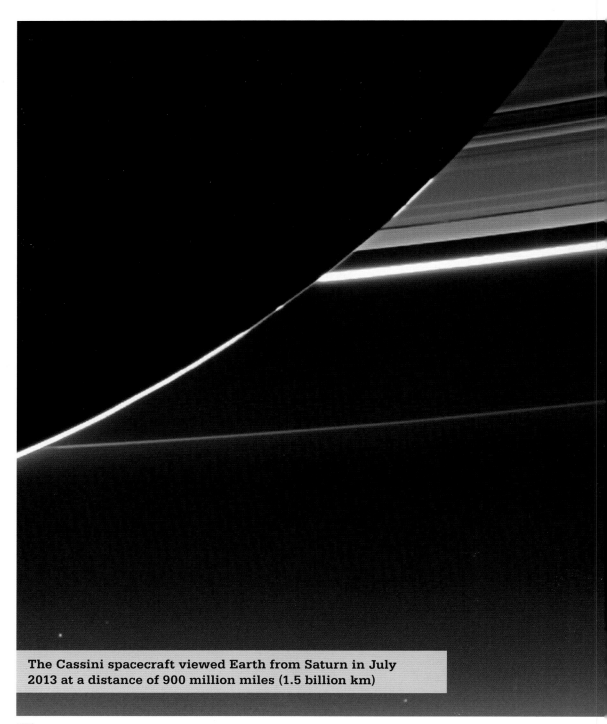

The Cassini spacecraft viewed Earth from Saturn in July 2013 at a distance of 900 million miles (1.5 billion km)

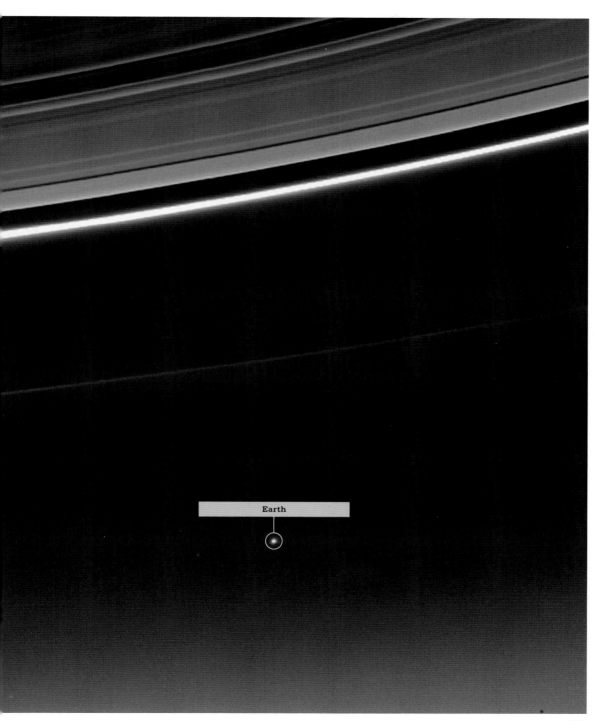

Earth

The **North Polar Sky**

This region of the celestial sphere is centered on the North Celestial Pole, which is situated close to the prominent star Polaris. From Europe and North America much of this area of sky is always above the horizon. But at midlatitudes in the Southern Hemisphere, such as Australia, most of it is permanently hidden from view.

Earth ....

North Polar sky

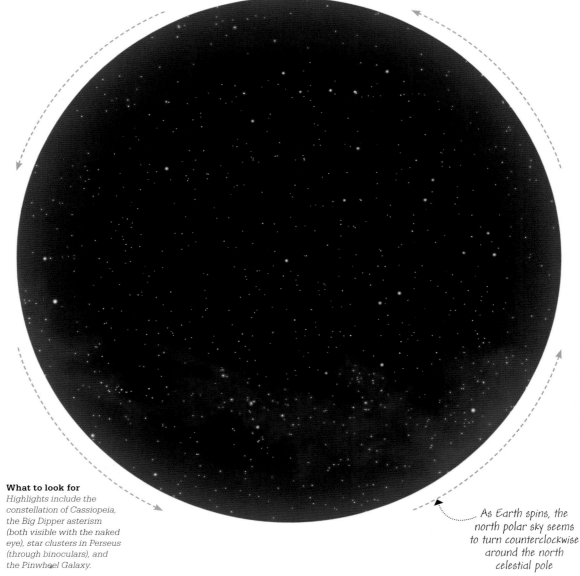

What to look for
Highlights include the constellation of Cassiopeia, the Big Dipper asterism (both visible with the naked eye), star clusters in Perseus (through binoculars), and the Pinwheel Galaxy.

.... As Earth spins, the north polar sky seems to turn counterclockwise around the north celestial pole

Key to the Star Map

Magnitudes

-1 0 1 2 3 4 5 6

Double stars

Variable stars

Open cluster

Globular cluster

Bright nebula

Galaxy

Supernova remnant

Planetary nebula

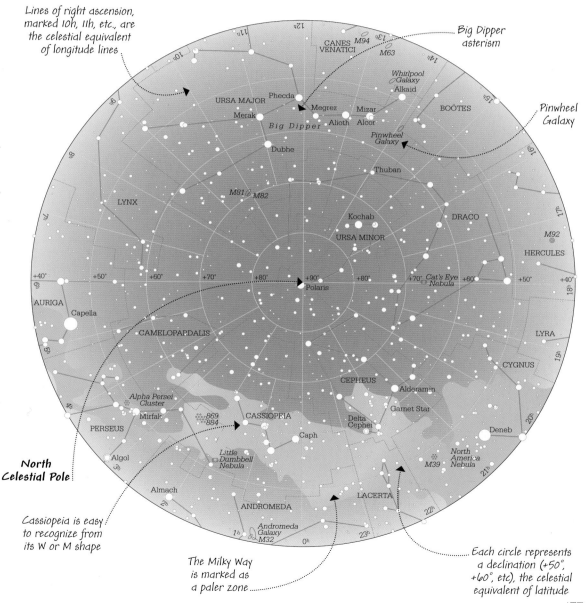

Lines of right ascension, marked 10h, 11h, etc., are the celestial equivalent of longitude lines

Big Dipper asterism

Pinwheel Galaxy

CANES VENATICI

M94

M63

Whirlpool Galaxy

Alkaid

URSA MAJOR

Phecda

Megrez

Mizar

BOÖTES

Merak

Big Dipper

Alioth

Alcor

Dubhe

Pinwheel Galaxy

Thuban

LYNX

M81

M82

Kochab

DRACO

M92

URSA MINOR

HERCULES

+40° +50° +60° +70° +80° +90° +80° +70° +60° +50° +40°

Polaris

Cat's Eye Nebula

AURIGA

Capella

LYRA

CAMELOPARDALIS

CYGNUS

CEPHEUS

Alderamin

Alpha Persei Cluster

Garnet Star

Deneb

Mirfak

869 884

CASSIOPEIA

Delta Cephei

North America Nebula

PERSEUS

Caph

M39

Algol

Little Dumbbell Nebula

North Celestial Pole

Almach

LACERTA

Cassiopeia is easy to recognize from its W or M shape

ANDROMEDA

Andromeda Galaxy M32

The Milky Way is marked as a paler zone

Each circle represents a declination (+50°, +60°, etc), the celestial equivalent of latitude

The **South Polar Sky**

This region of the celestial sphere is centered on the South Celestial Pole, which is situated in a fairly empty area of sky. From areas such as Australia, New Zealand, and southern Chile, much of this region of sky is always above the horizon. But from countries at midlatitudes in the Northern Hemisphere, most of it is permanently hidden from view.

Earth

South polar sky

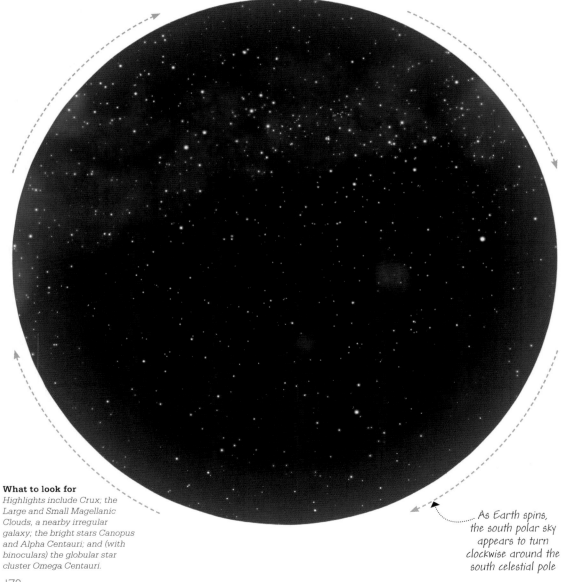

What to look for
Highlights include Crux; the Large and Small Magellanic Clouds, a nearby irregular galaxy; the bright stars Canopus and Alpha Centauri; and (with binoculars) the globular star cluster Omega Centauri.

As Earth spins, the south polar sky appears to turn clockwise around the south celestial pole

Key to the Star Map

Magnitudes

-1 0 1 2 3 4 5 6

Double stars

Variable stars

Open cluster

Globular cluster

Bright nebula

Galaxy

Supernova remnant

Planetary nebula

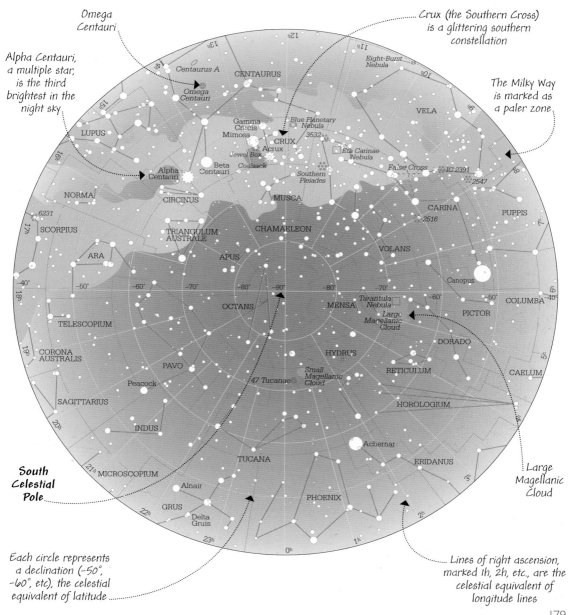

Omega
Centauri

Alpha Centauri,
a multiple star,
is the third
brightest in the
night sky

Crux (the Southern Cross)
is a glittering southern
constellation

The Milky Way
is marked as
a paler zone

South
Celestial
Pole

Each circle represents
a declination (–50°,
–60°, etc), the celestial
equivalent of latitude

Large
Magellanic
Cloud

Lines of right ascension,
marked 1h, 2h, etc., are the
celestial equivalent of
longitude lines

Centaurus A
Omega
Centauri
CENTAURUS
Eight-Burst
Nebula
VELA
Blue Planetary
Nebula
Gamma
Crucis
Mimosa
CRUX
Acrux
3532
Eta Carinae
Nebula
IC 2391
2547
LUPUS
Jewel Box
Coalsack
Southern
Pleiades
False Cross
Alpha
Centauri
Beta
Centauri
Southern
Pleiades
CARINA
2516
PUPPIS
NORMA
6231
CIRCINUS
MUSCA
COLUMBA
SCORPIUS
TRIANGULUM
AUSTRALE
CHAMAELEON
VOLANS
ARA
APUS
Canopus
–40°
–50°
–60°
–70°
–80°
–90°
–80°
–70°
–60°
50°
–40°
TELESCOPIUM
OCTANS
MENSA
Tarantula
Nebula
PICTOR
CORONA
AUSTRALIS
Large
Magellanic
Cloud
DORADO
CAELUM
PAVO
HYDRUS
RETICULUM
Peacock
Small
Magellanic
Cloud
SAGITTARIUS
47 Tucanae
HOROLOGIUM
INDUS
Achernar
MICROSCOPIUM
TUCANA
ERIDANUS
Alnair
GRUS
Delta
Gruis
PHOENIX

12h 11h 10h 9h 8h 7h 6h 5h 4h 3h 2h 1h 0h 23h 22h 21h 20h 19h 18h 17h 16h 15h 14h 13h

The **Equatorial Sky 1**

This part of the sky can be viewed during evenings from around September to November, looking to the south and overhead from the Northern Hemisphere or to the north (and overhead) from the Southern Hemisphere. Alternatively, it can be viewed in the early hours of the morning, looking in the same directions, from around June to August.

Earth ...... Equatorial sky 1

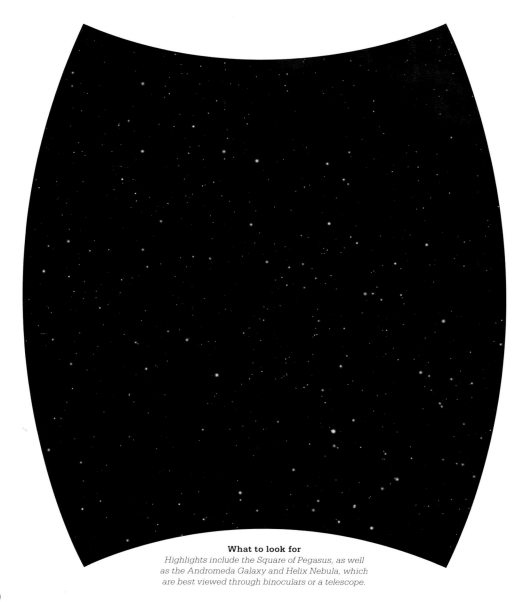

What to look for
Highlights include the Square of Pegasus, as well as the Andromeda Galaxy and Helix Nebula, which are best viewed through binoculars or a telescope.

Key to the Star Map

Magnitudes

-1 0 1 2 3 4 5 6

Double stars

Variable stars

Open cluster

Globular cluster

Bright nebula

Galaxy

Supernova remnant

Planetary nebula

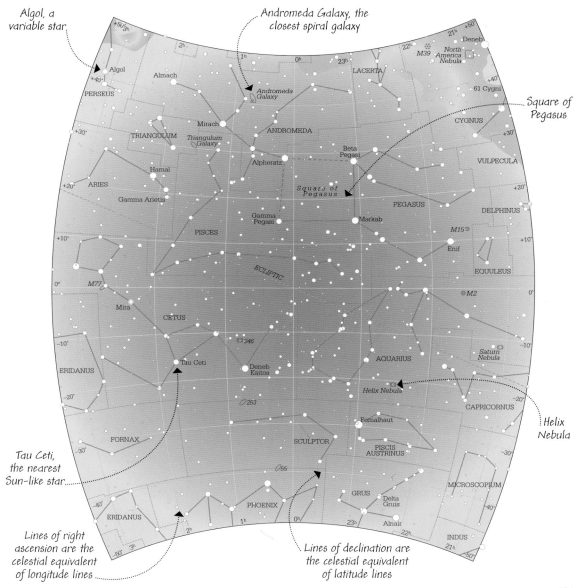

Algol, a variable star

Andromeda Galaxy, the closest spiral galaxy

Square of Pegasus

Helix Nebula

Tau Ceti, the nearest Sun-like star

Lines of right ascension are the celestial equivalent of longitude lines

Lines of declination are the celestial equivalent of latitude lines

Deneb
M39
North America Nebula
LACERTA
61 Cygni
CYGNUS
VULPECULA
DELPHINUS
M15
Enif
EQUULEUS
M2
Saturn Nebula
CAPRICORNUS
MICROSCOPIUM
Delta Gruis
GRUS
Alnair
INDUS
PISCIS AUSTRINUS
Fomalhaut
SCULPTOR
AQUARIUS
Helix Nebula
55
253
246
Deneb Kaitos
Tau Ceti
CETUS
Mira
M77
ERIDANUS
PHOENIX
FORNAX
ERIDANUS
PISCES
Gamma Pegasi
Markab
PEGASUS
Square of Pegasus
Beta Pegasi
Alpheratz
ANDROMEDA
Andromeda Galaxy
Mirach
TRIANGULUM
Triangulum Galaxy
Hamal
Gamma Arietis
ARIES
Almach
Algol
PERSEUS
ECLIPTIC

The **Equatorial Sky 2**

This sky region is best viewed on evenings from June to August, looking predominantly to the south from the Northern Hemisphere, or to the north from the Southern Hemisphere. It can also be viewed in the early hours of the morning from around March to May. The brightest part of the Milky Way dominates this part of the night sky.

Earth

Equatorial
sky 2

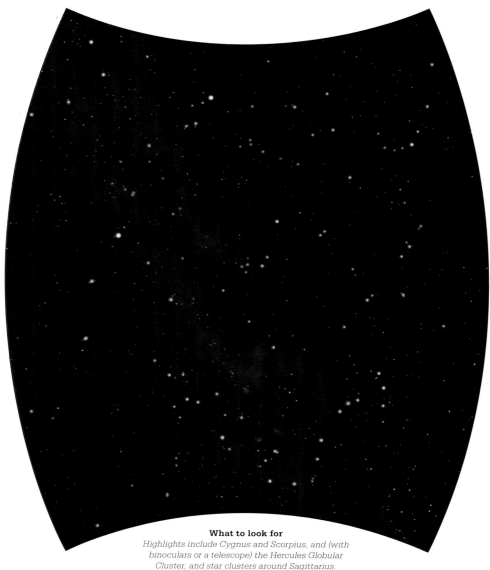

What to look for
Highlights include Cygnus and Scorpius, and (with binoculars or a telescope) the Hercules Globular Cluster, and star clusters around Sagittarius.

Key to the Star Map

Magnitudes

-1 0 1 2 3 4 5 6

Double stars   Open cluster   Bright nebula   Supernova remnant

Variable stars   Globular cluster   Galaxy   Planetary nebula

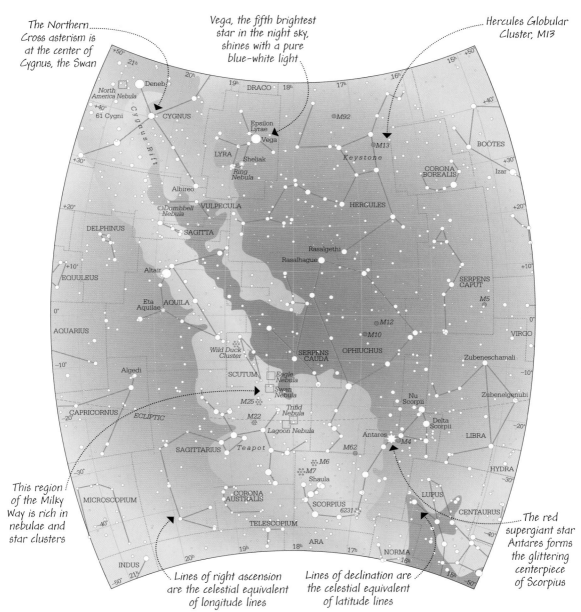

The Northern Cross asterism is at the center of Cygnus, the Swan

Vega, the fifth brightest star in the night sky, shines with a pure blue-white light

Hercules Globular Cluster, M13

North America Nebula

61 Cygni

Cygnus Rift

CYGNUS

Deneb

DRACO

M92

Epsilon Lyrae

Vega

LYRA

Sheliak

Keystone

M13

BOÖTES

CORONA BOREALIS

Izar

Ring Nebula

Albireo

VULPECULA

HERCULES

Dumbbell Nebula

DELPHINUS

SAGITTA

Rasalgethi

Rasalhague

EQUULEUS

Altair

SERPENS CAPUT

M5

Eta Aquilae

AQUILA

AQUARIUS

M12

M10

OPHIUCHUS

VIRGO

Wild Duck Cluster

SERPENS CAUDA

Zubeneschamali

Algedi

SCUTUM

Eagle Nebula

Zubenelgenubi

Swan Nebula

Nu Scorpii

CAPRICORNUS

ECLIPTIC

M25

Trifid Nebula

Delta Scorpii

LIBRA

M22

Lagoon Nebula

Antares

M4

SAGITTARIUS

Teapot

M62

HYDRA

MICROSCOPIUM

M6

M7

Shaula

LUPUS

CORONA AUSTRALIS

SCORPIUS

6231

CENTAURUS

TELESCOPIUM

ARA

NORMA

INDUS

This region of the Milky Way is rich in nebulae and star clusters

Lines of right ascension are the celestial equivalent of longitude lines

Lines of declination are the celestial equivalent of latitude lines

The red supergiant star Antares forms the glittering centerpiece of Scorpius

The **Equatorial Sky 3**

This area of sky can be seen during evenings from March to May, looking mainly to the south in the Northern Hemisphere, or to the north in the Southern Hemisphere. It can also be viewed in the hours after midnight from December to February. In this sky sector, there are only a few bright stars but many interesting deep-sky objects.

Earth

Equatorial sky 3

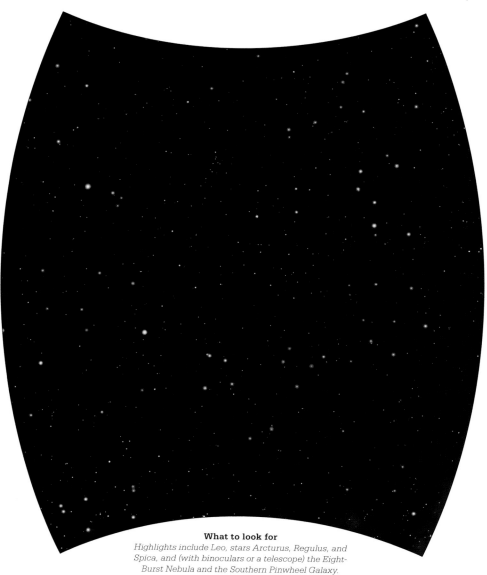

What to look for
Highlights include Leo, stars Arcturus, Regulus, and Spica, and (with binoculars or a telescope) the Eight-Burst Nebula and the Southern Pinwheel Galaxy.

Key to the Star Map

Magnitudes

-1 0 1 2 3 4 5 6

Double stars ✦
Variable stars ◉ ◎ ◦

Open cluster ✳
Globular cluster ⊛

Bright nebula ▢
Galaxy ◯

Supernova remnant ⊛
Planetary nebula ◐

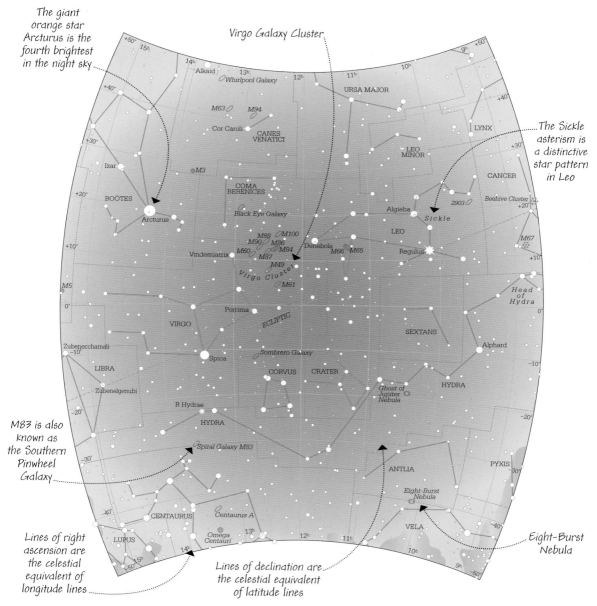

The giant orange star Arcturus is the fourth brightest in the night sky

Virgo Galaxy Cluster

The Sickle asterism is a distinctive star pattern in Leo

M83 is also known as the Southern Pinwheel Galaxy

Lines of right ascension are the celestial equivalent of longitude lines

Lines of declination are the celestial equivalent of latitude lines

Eight-Burst Nebula

Alkaid Whirlpool Galaxy URSA MAJOR
M63 M94
Cor Caroli CANES VENATICI
LYNX
LEO MINOR
Izar M3
CANCER
BOÖTES COMA BERENICES
Beehive Cluster
Arcturus Black Eye Galaxy
Algieba Sickle
2903
M67
M88 M100 LEO
M90 M86
Vindemiatrix M60 M84 Denebola M66 M65 Regulus
M87 M49
Virgo Cluster
M61
Head of Hydra
M5
Porrima ECLIPTIC
VIRGO
SEXTANS
Zubenecchamali Alphard
Spica Sombrero Galaxy
LIBRA CORVUS CRATER
Zubenelgenubi HYDRA
R Hydrae Ghost of Jupiter Nebula
HYDRA
Spiral Galaxy M83
PYXIS
ANTLIA
CENTAURUS Centaurus A
Eight-Burst Nebula
Omega Centauri VELA
LUPUS

The **Equatorial Sky 4**

This part of the sky is best viewed during evenings from December to February, looking predominantly to the south in the Northern Hemisphere, or to the north in the Southern Hemisphere. It is also visible in the early mornings from September to November. Many brilliant stars and prominent constellations feature in this sector.

Earth

Equatorial sky 4

What to look for

Highlights include Orion, Taurus, and Canis Major, and the Pleiades, and (through binoculars or a telescope) the Orion Nebula, Hyades, and Beehive clusters.

Key to the Star Map

Magnitudes

-1 0 1 2 3 4 5 6

Double stars

Variable stars

Open cluster

Globular cluster

Bright nebula

Galaxy

Supernova remnant

Planetary nebula

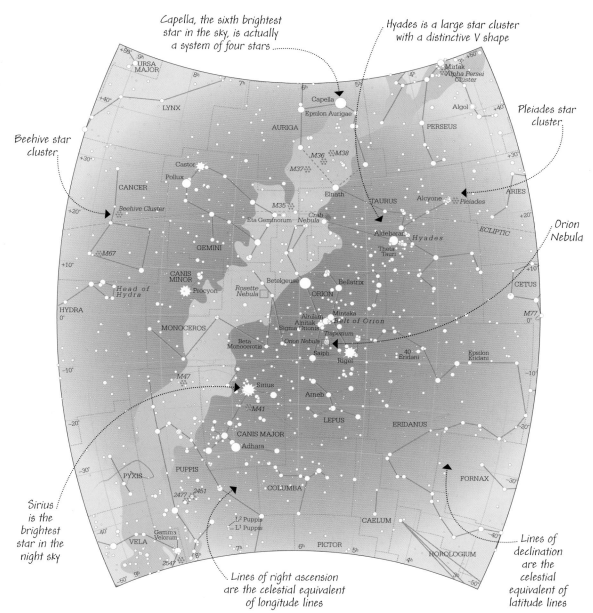

Capella, the sixth brightest star in the sky, is actually a system of four stars

Hyades is a large star cluster with a distinctive V shape

Pleiades star cluster

Beehive star cluster

Orion Nebula

Sirius is the brightest star in the night sky

Lines of right ascension are the celestial equivalent of longitude lines

Lines of declination are the celestial equivalent of latitude lines

URSA MAJOR
LYNX
AURIGA
Capella
Epsilon Aurigae
M36
M38
M37
Elnath
Mirfak
Alpha Persei Cluster
Algol
PERSEUS
ARIES
Alcyone
Pleiades
TAURUS
Aldebaran
Hyades
Theta Tauri
ECLIPTIC
Castor
Pollux
CANCER
Beehive Cluster
M35
Eta Geminorum
Crab Nebula
GEMINI
M67
CANIS MINOR
Head of Hydra
Procyon
HYDRA
Rosette Nebula
Betelgeuse
ORION
Bellatrix
CETUS
M77
MONOCEROS
Alnilam
Alnitak
Sigma Orionis
Mintaka
Belt of Orion
Trapezium
Beta Monocerotis
Orion Nebula
Saiph
Rigel
40 Eridani
Epsilon Eridani
M47
Sirius
M41
Arneb
LEPUS
ERIDANUS
CANIS MAJOR
Adhara
FORNAX
PYXIS
PUPPIS
2477
2451
COLUMBA
CAELUM
L² Puppis
L¹ Puppis
VELA
Gamma Velorum
PICTOR
HOROLOGIUM
2547

Index

Useful resources

Planispheres
DK Planisphere and Starfinder, Carole Stott and Giles Sparrow.
Latitude options: 55°N (UK); 35°N (US); 35°S (Australia)
Philip's Planisphere,
Latitude options: 23.5°N, 32°N, 42°N, 51.5°N, 35°S
The Night Sky Planisphere
David Chandler. Latitude options: 20–30°N, 30–40°N, 40–50°N, 50–60°N, 20–50°S
Rob Walrecht Planisphere
Latitude options: from 40°S to 66.5°N and several languages
Equatorial Guide to the Stars (Ken Press/Ken Graun)
Latitude options: 30°S-30°N

Desktop/Laptop Software
Starry Night Complete Space & Astronomy Pack
(Simulation Curriculum Corp)
Beginner level for PCs and AppleMacs
Redshift 7 Premium
(United Soft Media) Advanced level, for PCs only

iPhone and iPad "apps"
Stellarium Mobile (Noctua Software)
Pocket Universe (John Kennedy)
Android "apps"
Mobile Observatory (KreApp Development Software)
Sky Map (Google)
Star Walk (iTunes)

Books
The Practical Astronomer, Will Gater and Anton Vamplew (DK)
Stars and Planets Nature Guide, Robert Dinwiddie, Will Gater, Giles Sparrow, and Carole Stott (DK)
Eyewitness Companions: Astronomy, Ian Ridpath (DK)
Space, Carole Stott, Robert Dinwiddie, David Hughes, and Giles Sparrow (DK)
Universe, General Editor: Martin Rees (DK)
Unfolding our Universe Iain Nicolson (Cambridge University Press)

Online tools and calendars
NASA SkyWatch (for ISS, HST, and other satellites)
http://spaceflight.nasa.gov/realdata/sightings/SSapplications/Post/JavaSSOP/JavaSSOP.html
Astronomical Calendar night sky http://calsky.com/
Sea and sky
http://www.seasky.org/astronomy/astronomy-calendar-current.html
Heavens Above http://www.heavens-above.com
Sky View Café
Online planetarium.
http://www.skyviewcafe.com
Astronomy magazine
http://www.astronomy.com

About the Author

Robert Dinwiddie is a science writer, editor, and educator who specialises in astronomy, cosmology, Earth science, and the history of science. A keen traveller and amateur astronomer, he has worked in a variety of media, including books, magazines, and electronic publishing. This has included managing the development of the DK CD-ROM, *Eyewitness Encyclopedia of Space and the Universe*, co-authoring DK astronomy encyclopedias *Universe* and *Space*, the *Stars and Planets* nature guide, further DK science-based encyclopedias including *Earth*, *Science*, *Ocean*, and *Violent Earth*, and numerous other books.

Acknowledgments

Many people helped in the making of this book. DK would particularly like to thank:
Additional writing Ben Morgan; **Photography** Peter Anderson
Editorial assistance and proofreading May Corfield
Design assistance Prashant Kumar, Clare Marshall;
Illustrations Vanessa Hamilton; **Indexer** Jane Coulter

Picture credits

DK would like to recognise the efforts of all the enthusiastic astronomers and astro-photographers who share their images online. The publishers would like to thank the following for their kind permission to reproduce their photographs in this book: (Key: a-above; b-below/bottom; c-centre; f-far; l-left; r-right; t-top)

02 Courtesy Jim Misti. 4 NASA: Rocky Raybell (b). **6 ESO:** B.Tafreshi (twanight.org). **7 NASA:** JPL-Caltech/R. Gehrz (University of Minnesota) (cr). **9 NASA:** JPL-Caltech, SDSS (tr). **10 NASA:** Solar & Heliospheric Observatory (l). **13 Dorling Kindersley:** NASA/digitaleye © Jamie Marshall (tr). **14 NASA:** ESA, Buie/Southwest Research Institute (t). **18 NASA:** Rocky Raybell. **19 ESO:** Y. Beletsky (b). **NASA and The Hubble Heritage Team (AURA/STScI):** D. Jewitt (UCLA) (tr). **20 Fraser Gunn:** (tr). **NASA and The Hubble Heritage Team (AURA/STScI):** John Spencer (Lowell Observatory) (tl); L. Sromovsky and P. Fry (University of Wisconsin), H. Hammel (Space Science Institute),K. Rages (SETI Institute) (bl). **NASA:** E. Karkoschka (University of Arizona), and H.B. Hammel (Space Science Institute, Boulder, Colorado) (br). **21 NASA:** ESA, and The Hubble Heritage Team (STScI). **33 David Cortner. 42 © Australian Astronomical Observatory/David Malin. 44 Dan Heller Photography. Chris Picking:** (tl, tr). **50 NASA and The Hubble Heritage Team (AURA/STScI):** Hubble SM4 ERO Team. **51 NASA:** STScI Digitized Sky Survey/Noel Carboni (r). **52 NASA and The Hubble Heritage Team (AURA/STScI):** C.R. O'Dell and S.K. Wong (Rice University) (br). **54 ESO:** J.Perez (tl). **55 ESO. 56 NASA:** ESA, SSC, CXC, and STScI (l). **70 NASA:** JPL (br). **73 NASA:** JPL-Caltech/Harvard-Smithsonian CfA (b). **74 Amir Hossein Abolfath:** (l). **75 Jens Hackmann. 77 Arman Golestaneh. 79 NASA:** JPL-Caltech (b). **82–83 Peter Lawrence. 86 ESO:** F.Char (br). **87 Mark Humpage. 90–91 Ben Cooper/LaunchPhotography.com. 91 Rainbow Symphony Inc:** (tr). **92–93 NASA and The Hubble Heritage Team (AURA/STScI). 96 ESA/Hubble:** (br). **NASA:** (bl). **97 NASA:** JPL/USGS. **101 NASA and The Hubble Heritage Team (AURA/STScI):** (inset). **102 Jean-Charles Cuillandre/Canada-France-Hawaii Telescope Corporation:** (br). **NASA:** F. R. Ferraro (ESO/Bologna Obs.), M. Shara (STScI/AMNH), Hubble Heritage Team (AURA/STScI/NASA) (bl). **103 Jean-Charles Cuillandre/Canada-France-Hawaii Telescope Corporation.105 Courtesy Jim Misti:** (Inset). **NASA:** JPL-Caltech/UCLA (br). **107 Danny Lee Russell:** (t). **108 ESO:** Gendler, J.E. Ovaldsen & S. Guisard (b). **109 ESO:** (Inset). **112 NASA:** JPL/MSSS (br). **114 Stephen W. Ramsden/www.solarastronomy.org:** (bl). **115 NASA and The Hubble Heritage Team (AURA/STScI):** ESA, M. Robberto (STScIESA). **116 ESO:** (t, br). **NASA:** The Hubble Heritage Team (STScI) (bl). **117 Larry van Vleet. 119 ESO:** (Inset). **120 ESO:** S.Brunier (b). **121 ESO:** (t/Inset). **123 ESO:** INAF-VST/OmegaCAM. (t/Inset). **124 NASA. 127 David Davison:** (br). **128 Matt Kingsnorth:** (bl). **129 R Gendler:** Robert Gendler. **130 ESO:** (bl). **NASA:** Laurent Drissen, Jean-Rene Roy and Carmelle Robert (Department de Physique and Observatoire du mont Megantic, Universite Laval) (br); ESA and The Hubble Heritage Team (STScI/AURA) (tl); ESA, and The Hubble Heritage Team (STScIAURA) (tr). **131 NASA:** ESA/John Bahcall (Institute for Advanced Study, Princeton) Mike Disney (University of Wales) (b); The Hubble Heritage Team (STScI) (t). **133 David Ratledge:** (t/Inset). **134 Justin Whitten:** (b). **135 NASA:** JPL-Caltech (t/Inset). **137 ESO:** B.Tafreshi (twanight.org). **138 NASA:** Courtesy of Science/AAAS. **140 Fred Espenak. 141 Philipp Salzgeber:** (b). Doug Zubenel: (t). **142 NASA:** Astronomy Education Services/ Gingin Observatory (bl). **143 ESO:** S. Deiries. **144-145 NASA:** JPL-Caltech/UCLA. **150 NASA:** JPL-Caltech/Space Science Institute (br). **153 Fraser Gunn:** (tr). **NASA:** NASA Planetary Photojournal (br). **155 Photo courtesy of Bryan Bradley, Babylon, NY, USA:** (cr/Inset). **NASA:** D. Williams, N. A. Sharp, AURA, NOAO, NSF (tl/Inset). **156 NASA and The Hubble Heritage Team (AURA/STScI):** (bl). **Keith Quattrochi:** (br). **157 Gemini Observatory:** AURA. **158 NASA:** The Hubble Heritage Team (STScI) (b). **159 ESO:** (cr/Inset). NASA: The Hubble Heritage Team (STScI)/AURA/ESA) (t/Inset). **161 NASA:** Solar & Heliospheric Observatory (br). **162 NASA:** (tl, cr). **163 NASA:** (bl, r). **165 NASA and The Hubble Heritage Team (AURA/STScI):** (cr/Inset). **167 NASA:** JPL-Caltech (br). Robert Vanderbei (t). **169 ESO:** (cr/Inset). **170 Chandra X-Ray Observatory:** NOAO/CTIO (cl). **171 NASA:** JPL-Caltech/R. Gehrz (University of Minnesota). **172 © Australian Astronomical Observatory/ David Malin:** (bl). **NASA:** JPL-Caltech/UCLA (br). 173 Chandra X-Ray Observatory: NASA/CXC/MIT/UMass Amherst/M.D.Stage et al. **174-175 NASA:** JPL-Caltech/Space Science Institute.

Every effort has been made to trace the copyright holders. The publishers apologize for any unintentional omissions and would be pleased, in such cases, to place an acknowledgment in future editions of this book.

All other images © Dorling Kindersley
For further information see: **www.dkimages.com**